W9-BFY-796
3 1611 00326 2331

# The Intended Mathematics Curriculum as Represented in State-Level Curriculum Standards

## Consensus or Confusion?

a volume in
**Research in Mathematics Education Series**

Series Editor:
Barbara J. Dougherty, The University of Mississippi

## Research in Mathematics Education Series

Barbara J. Dougherty, Series Editor

*The Intended Mathematics Curriculum as Represented in State-Level Curriculum Standards: Consensus or Confusion?* (2006)
edited by Barbara J. Reys

# The Intended Mathematics Curriculum as Represented in State-Level Curriculum Standards

## Consensus or Confusion?

*edited by*

**Barbara J. Reys**
***University of Missouri-Columbia***

Charlotte, North Carolina • www.infoagepub.com

Library of Congress Cataloging-in-Publication Data

The intended mathematics curriculum as represented in state-level curriculum standards : consensus or confusion? / edited by Barbara Reys.
p. cm. — (Research in mathematics education)
Includes bibliographical references.
ISBN 1-930608-52-7 (pbk.) — ISBN 1-930608-53-5 (hardcover)
1. Mathematics—Study and teaching—United States. 2. Education—Standards—United States. 3. Curriculum evaluation—United States. I. Reys, Barbara.
QA13.I546 2006
510.71'073—dc22

2006033028

ISBN 13: 978-1-930608-52-8 (pbk.)
ISBN 13: 978-1-930608-53-5 (hardcover)
ISBN 10: 1-930608-52-7 (pbk.)
ISBN 10: 1-930608-53-5 (hardcover)



Printed in the United States of America

# CONTENTS

# ACKNOWLEDGMENT

This report is based on the work of the Center for the Study of Mathematics Curriculum, supported by the National Science Foundation under Grant No. ESI-0333879. Any opinions, findings, and conclusions or recommendations expressed in this material are those of the author(s) and do not necessarily reflect the views of the National Science Foundation.

# PREFACE

Since 2001, many states have developed new, more specific mathematics curriculum frameworks outlining the intended curriculum, K–8. While some of these documents are intended to be "models" for districts to utilize in shaping local curriculum specifications, others are mandatory, specifying the mathematics all students within the state are expected to learn at particular grades. All appear to serve as guidelines for shaping annual state-wide grade level assessments. As a collection, the new state mathematics curriculum standards represent the mathematics students in the United States are expected to learn.

In developing the newest version of curriculum standards, many states provide increased levels of specificity over previous standards, in part due to No Child Left Behind (NCLB) requirements related to specification of performance standards and accompanying annual assessments in Grades 3–8. While local control of educational decisions, including curriculum standards, is a hallmark of American education, increased accountability has focused more attention on state-level curriculum decisions. A recent survey indicates that the state-level curriculum documents are receiving as much, if not more, attention by school administrators and teachers as the textbooks purchased to support curriculum implementation (Reys, Dingman, Sutter, & Teuscher, 2005).

Given the higher profile of state-level curriculum standards documents, the Center for the Study of Mathematics Curriculum (CSMC), a National Science Foundation–funded Center for Learning and Teaching, set out to describe the level of consistency in learning goals across these documents. That is, to what extent are particular learning goals emphasized within state documents and what is the range of grade levels where these learning goals are emphasized?

This volume represents the first detailed analysis of the grade placement of particular learning goals across all state-level curriculum documents published and current as of May 2005. One of the difficulties of this task was determining the intent of the learning expectations across states. Due to the vagueness of some learning expectations as well as different terminology used across state documents, interpretations were made that may not reflect the intent of the document. For any misinterpretation the authors of this report assume full responsibility.

The report documents the current situation regarding grade-level mathematics curriculum specification in the United States and highlights a general lack of consensus across states. As states continue to work to improve learning opportunities for all students, we hope this report serves as a useful summary to inform future curriculum decisions. We also hope the report stimulates discussion at the national level regarding roles and responsibilities of national agencies and professional organizations with regard to curriculum leadership. We believe that serious and collaborative work that results from such a discussion can contribute to a solution to the "mile wide and inch deep" U. S. curriculum.

# EXECUTIVE SUMMARY

Since the passage of the federal No Child Left Behind Act (NCLB, 2001) state departments of education and local school districts have been scrambling to address the law's requirements. One major area of focus has been identification of student learning expectations in mathematics. These learning expectations, sometimes called curriculum standards, are referred to in recent state documents as grade-level learning expectations (GLEs). They convey the specific mathematics content that students at particular grades are expected to learn (and teachers are expected to teach).

NCLB requires that states adopt "challenging academic content standards" in mathematics, reading/language arts, and science that (1) specify what children are expected to know and be able to do; (2) contain coherent and rigorous content; and (3) encourage the teaching of advanced skills (NCLB, 2001). Furthermore, states are required, beginning no later than school year 2005–2006, to measure the achievement of students against the state standards in Grades 3 through 8. In fact, 39 states (the District of Columbia and the Department of Defense Education Activity are counted as states) have published new mathematics curriculum standards documents since 2002 (see Table 1). These new documents include learning expectations organized by grade for most, if not all, of the Grades K–8. The current set of state-level mathematics standards documents, including those that articulate grade-level learning expectations

**Table 1. Publication Dates of Most Recent State-Level Mathematics Curriculum Documents (as of 2/1/06)**

| *Year* | *Number* | *States* |
|---|---|---|
| 2006 | 1 | MS |
| 2005 | 10 | AK, CA, CT, DC, HI, ID*, NV*, NY, ND, TX |
| 2004 | 15 | AR, DoDEA, GA, KY*, LA, ME, MD, MA, MI, MO, NH*, RI*, SD, VT, WA |
| 2003 | 8 | AL, AZ, KS, MN, NC, UT, WV, WY |
| 2002 | 5 | NJ, NM, OK, OR, VA |
| 2001 | 3 | OH, SC, TN |
| 2000 | 2 | IN, NE |
| Pre-2000 | 7 | CO, DE, FL, IL, MT, PA, WI |
| None | 1 | IA |
| TOTAL | 52 | |

*Note:* * Draft document

(GLE) or secondary course-level learning expectations (CLE), can be found at http://mathcurriculumcenter.org/states.php

The GLE documents represent the intended curriculum within the respective state. However, the extent to which these documents present similar messages about content emphasis and grade placement is unclear. The purpose of this study was to describe the emphasis and grade-level placement of particular learning expectations as presented in state GLE documents and to document variations across states. It does not provide a comprehensive summary of the documents. Rather, attention to particular mathematical topics or themes in three areas (Number and Operation, Algebra, and Reasoning) was the focus of the study.

This report describes the amount of variation regarding specified grades at which states call for particular learning goals/expectations. That is, we examined the extent to which there is consensus across state documents on when students should study particular topics. We examined only state mathematics standards documents that included elementary and middle school grade-by-grade learning expectations including at least Grades 3–7—42 in all at the time of the study (see Table 2).

The extent to which the content emphasized at various grade levels is the same or different has implications for the development of publisher-generated textbooks, teacher preparation, and comparisons of student performance. Described here (and reported in more depth in the full report) are major findings of the analysis of three areas (Number and Operation, Algebra, and Reasoning) across elementary and middle school state GLE documents.

**Table 2. Organization of K–8 Mathematics Grade-Level Learning Expectations (GLE) Documents by State and Grade Level (as of 2/1/06)**

| *Elementary/Middle School Learning Expectations* | | | |
|---|---|---|---|
| *GLE Documents (Grades K–8)* | *GLE Documents (Other Grades)* | *Grade-Band Documents* | *No GLE or Grade-Band Documents* |
| AL, AZ, AR, CT, DoDEA, DC, FL, GA, HI, ID, IN, KS, LA, MD, MI, MN, MS, MO, NV, NH, NM, NY, NC, ND, OH, OK, OR, RI, SC, SD, TN, TX, VT, VA, WA, WV, WY | AK (3–10)<br>CA (K–7)<br>ME (3–8)<br>NJ (3–8)<br>UT (K–7)<br>KY (4–8)[2] | CO[1] (K–4, 5–8, 9–12)<br>DE (K–3, 4–5, 6–8, 9–10)<br>IL[1] (early elem., late elem., middle/junior high, early HS, late HS)<br>MA (1–2, 3, 3–4, 5, 5–6, 7, 7–8)<br>MT (K–4, 5–8, 9–12)<br>NE (K–1, 2–4, 5–8, 9–12)<br>PA[1] (K–3, 4–5, 6–8, 9–10)<br>WI[1] (K–4, 5–8, 9–12) | IA |
| 37 | 6 | 8 | 1 |
| | 51 | | 1 |

*Note:* [1] CO, IL, PA, and WI have Assessment Frameworks dated 2003, 2004, or 2005 (CO [2003]: 3–10, IL [2004]: Grades 3–8, PA [2004]: Grades 3–8 and 11, WI [2005]: Grades 3–8 and 10)
[2] Since the KY document does not include Grade 3 GLEs, we did not include it in our analysis.

## FINDINGS REGARDING NUMBER AND OPERATION STRAND

Learning expectations related to the Number and Operation strand account for about a third of the total number of GLEs across all the K–8 state documents and emphasis on this strand is most prominent in Grades K–5. Within the strand, topics identified for analysis include fluency with basic number combinations (basic facts), multidigit whole number and fraction computation, estimation, and messages related to the role of calculators as computational tools. A summary of major findings follows.[1]

### Basic Number Combinations

The term basic number combinations refers to the set of single-digit combinations ($1 + 1, 1 + 2, \dots 9 + 9; 1 \times 1, 1 \times 2, \dots 9 \times 9$) whose sum (or product) students are expected to recall efficiently and accurately. Table 3 summarizes the grade at which 39 state documents (those that include at least Grades K–6) indicate that basic number combination fluency is expected. The most common grade placement for fluency with both addition and subtraction combinations is Grade 2. The most common grade placement for

**Table 3. Grade Placement of Learning Expectations Related to Fluency With Basic Number Combinations for Each Operation***

| *Operation* | *Grade* | *Number of States* | *Operation* | *Grade* | *Number of States* |
|---|---|---|---|---|---|
| Addition | 1 | 8 | Subtraction | 1 | 7 |
| | 2 | 28 | | 2 | 27 |
| | 3 | 2 | | 3 | 3 |
| | Not specified | 1 | | Not specified | 2 |
| Multiplication | 3 | 13 | Division3 | 3 | 6 |
| | 4 | 22 | | 4 | 20 |
| | 5 | 1 | | 5 | 3 |
| | 6 | 1 | | 6 | 1 |
| | Not specified | 2 | | Not specified | 9 |

*Note:* *For this analysis 39 of the 42 state documents were included (those that covered at least K–6).

**Table 4. Grade Placement of Culminating Learning Expectations Related to Fluency With Whole Number Computation for Each Operation**

| *Operation* | *Grade* | *Number of States* | *Operation* | *Grade* | *Number of States* |
|---|---|---|---|---|---|
| Addition | 1 | 1 | Subtraction | 1 | 1 |
| | 2 | 3 | | 2 | 2 |
| | 3 | 14 | | 3 | 15 |
| | 4 | 15 | | 4 | 15 |
| | 5 | 5 | | 5 | 5 |
| | 6 | 3 | | 6 | 3 |
| | Not specified | 1 | | Not specified | 1 |
| Multiplication | 3 | 2 | Division | 3 | 0 |
| | 4 | 21 | | 4 | 12 |
| | 5 | 15 | | 5 | 23 |
| | 6 | 3 | | 6 | 6 |
| | Not specified | 1 | | Not specified | 1 |

multiplication and division combinations is Grade 4. Note that the range in grade levels where fluency is expected is 2–3 years for each operation.

## Multidigit Whole Number Computation

The grade at which students are introduced to multidigit whole number computation and the grade at which fluency (proficiency with efficient and accurate methods) is expected varies considerably across the state GLE documents. For example, in some states students begin adding multidigit numbers as early as kindergarten while in other states this work begins in Grade 3. Table 4 summarizes the grade at which students are expected to be fluent with multidigit whole number computation for each operation. Forty-two state documents were reviewed for this analysis (those that include at least Grades 3–7). As noted, the culminating GLE (where fluency is expected) for addition of multidigit whole numbers ranges from Grade 1 to Grade 6 across the state documents. Multidigit multiplication is typically a focus at Grades 3 or 4 with fluency expected one year later (in Grades 4 or 5). Multidigit whole number division begins as early as Grade 3 in some states with an expectation of fluency most typically at Grade 5.

When particular learning expectations are examined, further variation is evident. For example, some state documents specify that students should be fluent in adding two- or three-digit numbers and others specify very large numbers (one state specifies computational fluency with nine-digit numbers).

## Fraction Computation

Attention to fractions within the school mathematics curriculum spans the full K–8 continuum and includes the introduction and development of the concept of a fraction; multiple representations of fractions; equivalence of fractions; conversions among fraction, decimal, and percent forms; and computation with fractions. As with whole number computation, state documents differ in their trajectory regarding the development of computational fluency with fractions. Table 5 provides a summary of the grade level at which states introduce computation with fractions. Table 6 summarizes the grade level where students are expected to be fluent computing with fractions. Once again, expectations span several years and highlight lack of consensus among states.

## Role of Calculators

A recent report published by the Thomas B. Fordham Foundation (Klein, Braams, Parker, Quirk, Schmid & Wilson, 2005) criticized state standards documents for their "overreliance" on calculators. Our review of the state GLE documents does not support this finding. We examined 42 state GLE documents, compiling each learning expectation that included one or both terms, "calculator" and/or "technology." Eleven of

**Table 5. Number of States and Grade Level When State GLE Documents *Introduce* Computation With Fractions by Operation**

| *Grade* | *Addition and Subtraction of Fractions* | *Multiplication of Fractions* | *Division of Fractions* |
|---|---|---|---|
| 1 | 2 States | | |
| 2 | | | |
| 3 | 7 States | | |
| 4 | 22 States | 1 State | 1 State |
| 5 | 9 States | 10 States | 6 States |
| 6 | 1 State | 25 States | 27 States |
| 7 | 1 State | 5 States | 6 State |
| 8 | | 1 State | 1 State |
| Not specified | | | 1 State |

**Table 6. Number of States and Grade Level When State GLE Documents Indicate *Expectation of Fluency With* Addition, Subtraction, Multiplication, and Division of Fractions**

| *Grade* | *Addition and Subtraction of Fractions* | *Multiplication of Fractions* | *Division of Fractions* |
|---|---|---|---|
| 4 | 1 State | | |
| 5 | 15 States | 2 States | 1 State |
| 6 | 20 States | 25 States | 24 States |
| 7 | 6 States | 13 States | 14 States |
| 8 | | 1 State | 1 State |

the 42 state documents make no mention of either term within the set of learning expectations. Another 18 of 42 state documents include 10 or fewer references to calculators/technology. The mean number of GLEs referencing calculators in the 31 state documents that do reference either term is 12.8 per state (1.42 per grade).

In the 31 documents that reference one or both terms, we identified a total of 430 learning expectations (less than 3% of the total number of learning expectations) utilizing either term. After eliminating GLEs that referred specifically to computer software (34 in all), a total of 396 GLEs were coded for analysis. As might be expected, the number of GLEs referring to "technology" or "calculators" increases as the grades increase from K–8 (see Table 7). As noted, the largest concentration of references to calculators/technology is in the middle grades. In fact, 211 of the 396 (53%) calculator-related GLEs identified are found at Grades 6, 7, or 8.

In addition to counting the number of references, we coded the implied or stated role or purpose of calculator/technology within the GLEs. At the

**Table 7. References to "Calculators" or "Technology" Within Learning Expectations by Grade Level Across 31 State GLE Documents (Those Which Include at Least one Reference to These Terms)**

| *Grade* | *Total Number of References Across All Documents* | *Mean Number of References per Document* |
|---|---|---|
| K | 8 | 0.26 |
| 1 | 20 | 0.65 |
| 2 | 27 | 0.87 |
| 3 | 36 | 1.16 |
| 4 | 44 | 1.42 |
| 5 | 50 | 1.61 |
| 6 | 59 | 1.90 |
| 7 | 66 | 2.13 |
| 8 | 86 | 2.77 |
| Total | 396 | 1.42 |

K–2 level, emphasis is on using tools (calculator or technology) to develop or demonstrate conceptual understanding. For Grades 3–5 the most common role is for developing concepts and/or solving problems. For Grades 6–8 the most common role of the calculator/technology specified in the state documents is to solve problems and/or display data.

In summary, attention to calculators and technology in all but a few state documents is limited and focused on use as tools for conceptual development and problem solving rather than as an alternative to computational fluency. In fact, all of the documents referring to calculators/technology are explicit in emphasizing that these tools do not replace the need for computational fluency.

## FINDINGS REGARDING ALGEBRA STRAND

Within the K–8 Algebra strand, five general categories of GLEs that accounted for approximately 90% of the learning expectations were identified: Patterns; Functions; Equations, Expressions and Inequalities (EEI); Properties; and Relationships Between Operations.

Figure 1 shows the total number of algebra expectations in three categories (Patterns, Functions, and EEI) that account for the greatest proportion of GLEs in the Algebra strand (about 80%). The graph shows that emphasis begins in kindergarten and steadily increases over Grades K to 5 followed by a more dramatic increase in Grades 6 to 8. Figure 2 shows the number of expectations for each substrand separately. When the three areas of Patterns, Functions, and EEI are graphed on the same axes, the

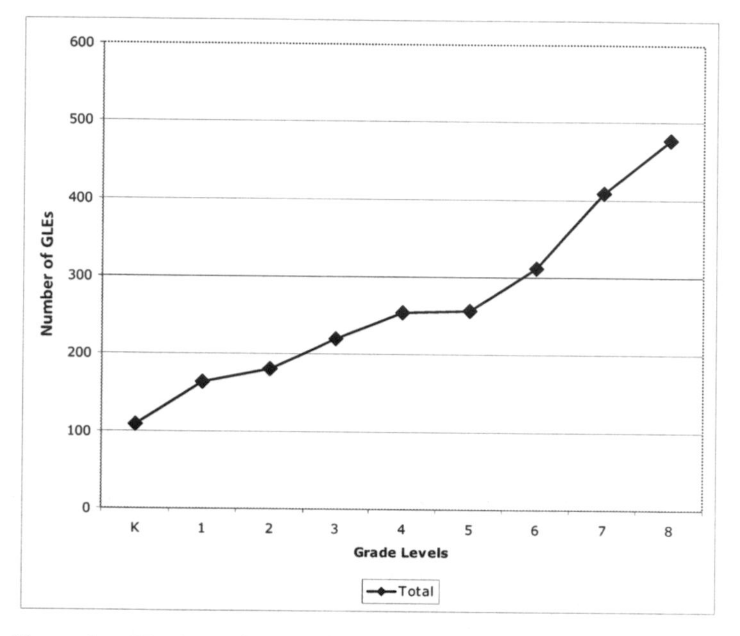


Figure 1. Total number of GLEs in Patterns, Functions, and EEI across grade levels.

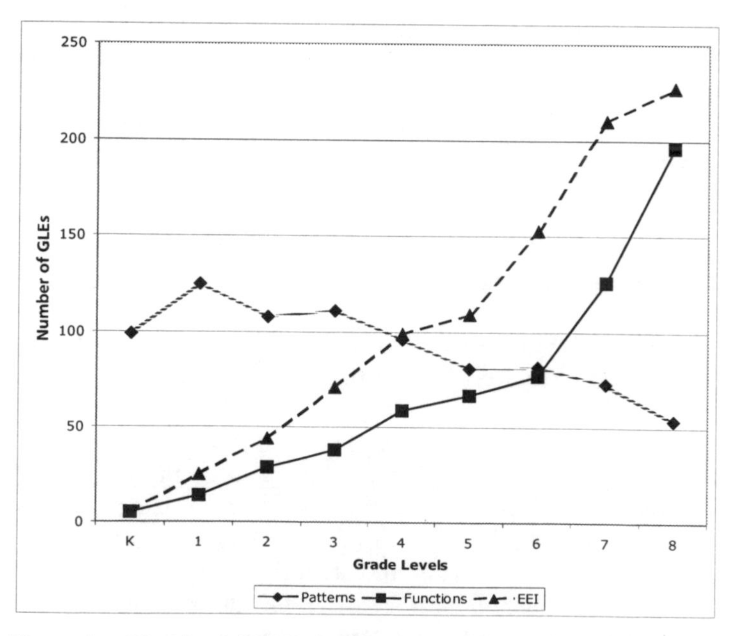


Figure 2. Number of GLEs in Patterns, Functions, and EEI across grade levels.

dominance of Pattern GLEs in Grades K to 3 with a steady decline over Grades 4 to 8 is apparent. The topics of Function and EEI steadily increase in emphasis (as judged by the number of GLEs) from Grades K–4 with dramatic increases in EEI from Grades 5 to 8 and Functions from Grades 6 to 8. Overall, the emphasis on EEI is predominant over both Functions and Patterns in Grades 4 to 8.

The number of learning expectations focused on Functions and EEI is an indication of the nature of algebra in the articulated school mathematics curriculum. The EEI strand represents what might be called "symbolic algebra," which suggests that algebra, particularly in the later grades, is focused on the development of symbolic algebra, or an equation-solving-driven algebra, more than on a function-based algebra.

## Algebra Curriculum for Grades K–8

In algebra few expectations reach mastery over Grades K–8. However, there is ample evidence that states vary substantially in the grade levels at which they concentrate on particular algebra topics. For example, the levels at which states expect the commutative property of multiplication to be taught vary from Grade 2 to 8 with Grades 3 and 4 having the greatest concentration of states. The levels at which states expect knowledge of variables ranges from kindergarten to Grade 8 with the greatest emphasis in Grades 4–7.

In order to have a metric that would represent a minimal level of agreement, we took 21 states (half the 42 state documents analyzed) as our benchmark. When we held this standard for the "common" K–8 algebra curriculum, very few topics made the cut (see Table 8). Table 8 does not tell the whole story, but it does give a picture of the core algebra concepts on which at least 21 states agree should be taught somewhere in Grades K–8.

While state standards documents include learning expectations related to algebra concepts in lower grade levels, the migration is not as apparent as the rhetoric in the United States would imply. There is a gradual buildup to more symbolic algebra at Grades 7 and 8, but the work at the lower grades seems to be more conceptual with gradual exposure to ideas. This analysis shows that there is a core of agreement on topics included in K–8 among at least half the states. However, there appears to be little overall agreement across documents in the algebra expectations for a particular grade level. In fact, there were no concepts or topics in algebra for which all 42 states at a given grade level include an expectation specific enough to code for the concept or topic. The greatest agreement reflected in our analysis is that 39 of the 42 documents include an expec-

**Table 8. Algebra Topics/Concepts in At Least 21 of 42 State Documents Analyzed**

| | | |
|---|---|---|
| Patterns | Classification of objects<br>Sorting of objects<br>Rule/generalization<br>Growing and shrinking patterns<br>Patterns involving skip counting | Repeating patterns<br>Numeric patterns<br>Geometric figure/shape patterns<br>Sequences |
| Functions | Rule/generalization<br>Change<br>Independent/dependent variables | Linear functions<br>Slope<br>Nonlinear functions |
| Expressions, Equations, and Inequalities | Variables<br>Expressions<br>Formulae<br>Number sentences/equations | 1-Step equations<br>2-Step equations<br>Inequalities |
| Properties | Commutative property of addition<br>Commutative property of multiplication<br>Associative property of addition<br>Associative property of multiplication | Distributive property<br>Additive identity<br>Multiplicative identity<br>Inverse (additive and multiplicative) |
| Relationships Between Operations | Addition and subtraction as inverse operations<br>Multiplication and division as inverse operations | Multiplication as repeated addition<br>Division as repeated subtraction<br>Order of operations |

tation that students should study algebraic expressions in Grade 7. The next highest level of agreement is that 32 documents include the study of variables at Grade 5 and expressions at Grade 8. The major result from our analysis is the *lack of agreement* on what should be expected at each grade level in the substrands of algebra.

## FINDINGS REGARDING REASONING STRAND

The importance of reasoning is clearly recognized as a K–8 learning goal based on a review of the state curriculum documents. In some state documents a "reasoning" strand provides the organizational structure for conveying intended emphasis on reasoning. Other state documents weave goals related to the development of reasoning throughout the content strands. However, there appears to be no consistency across state documents related to emphasis on reasoning at particular grade levels.

The major emphasis of our analysis focused on learning expectations pertaining to reasoning for verification. Learning expectations related to verification were identified then coded into categories as noted in Table 9.

**Table 9. Number of State Standards Documents, by Grade, That Include GLEs in Each Category of the Reasoning Strand**

| *Reasoning Focus* | *K* | *1* | *2* | *3* | *4* | *5* | *6* | *7* | *8* |
|---|---|---|---|---|---|---|---|---|---|
| Prediction | 8 | 17 | 24 | 24 | 24 | 26 | 22 | 25 | 27 |
| Generalization | 2 | 1 | 8 | 5 | 9 | 10 | 10 | 12 | 12 |
| Verification | 2 | 1 | 2 | 5 | 7 | 7 | 6 | 6 | 13 |
| Justification | 1 | 1 | 8 | 12 | 14 | 23 | 20 | 19 | 24 |
| Conclusion/Inference | 1 | 6 | 9 | 12 | 13 | 16 | 15 | 16 | 17 |
| Conjecture | 0 | 0 | 1 | 2 | 5 | 7 | 6 | 13 | 10 |
| Testing | 1 | 1 | 4 | 6 | 12 | 10 | 6 | 9 | 7 |
| Argument | 0 | 0 | 1 | 0 | 2 | 6 | 3 | 7 | 11 |
| Evaluation | 0 | 0 | 3 | 2 | 2 | 7 | 9 | 9 | 14 |

The majority of these learning expectations were found in three content strands: Data Analysis/Probability, Algebra, and Geometry. Table 9 summarizes the number of state standards documents that address one or more of these topics by grade. For example, 17 state documents include at least one GLE related to prediction in Grade 1. As noted, prediction is a common theme across grade levels and all categories of reasoning receive greater emphasis in Grades 4–8 than in K–3.

In general, we find that reasoning is not well articulated or integrated across K–8 standards documents. When reasoning GLEs are organized within a separate strand they tend to be broad and general, and isolated from specific content.

In summary, most state standards documents give attention to reasoning, incorporating learning expectations related to reasoning either within a separate, designated strand or by weaving messages about reasoning throughout the content strands. However, most state standards fail to address reasoning aspects in a thorough and comprehensive manner across grade levels and content strands. In addition, clarity and specificity of reasoning learning expectations vary across and within state documents.

## RECOMMENDATIONS REGARDING SPECIFICATION OF LEARNING EXPECTATIONS

Findings from this study confirm that state mathematics curriculum documents vary along several dimensions including grain size (level of specificity of learning expectations), language used to convey learning outcomes (understand, explore, memorize, and so on), and the grade placement of particular learning expectations. We offer some suggestions for groups that engage in future efforts to specify grade-level mathematics learning expectations.

- *Identify major goals or focal points at each grade level, K–8.* At each grade, we recommend a general statement of major goals for the grade. These general goals may specify emphasis on a few strands of mathematics or a few topics within strands. These general goals should be coordinated across all grades, K–8, to ensure curricular coherence and comprehensiveness. Offering these major goals will provide guidance to teachers in appropriation of instructional time. It may also help reduce superficial treatment of many mathematical topics, a common criticism of the United States mathematics curriculum. For guidance in determining the major goals or focal points, we suggest collaboration across states and/or guidance from groups such as NCTM and Achieve, Inc.
- *Limit the number of grade-level learning expectations to focus instruction and deepen learning.* The set of learning expectations per grade level should be manageable given the school year. Along with the statement of general goals and priorities for a particular grade, we suggest that the set of learning expectations per grade be limited to 20–25. This number is similar to curriculum standards documents in other countries and may help authors of standards documents develop an appropriate grain size for communicating learning expectations.
- *Organize K–8 grade-level learning expectations by strand.* We recommend that mathematics curriculum standards be organized by grade and by content strand. Further, we recommend that attention be given to both content strands (e.g., Number and Operation, Geometry, Measurement, Algebra, Data Analysis and Probability) and important mathematical processes (e.g. Problem Solving, Reasoning, and Representations).
- *Develop clear statements of learning expectations focusing on mathematics content to be learned.* We recommend that learning expectations be expressed succinctly, coherently, and with optimum brevity, limiting the use of educational terms (jargon) that may not communicate clearly to the intended audience of teachers, school leaders, and parents. Learning expectations should focus on the mathematics to be learned rather than pedagogy to be employed in presenting the mathematics. The set of learning expectations for a grade should include mathematics to be learned at that grade level (not just what will be assessed).

- *Limit the use of examples within learning expectations*. Some state GLE documents include examples (occasionally or frequently, depending on the document) to clarify the learning expectations and others do not. In some documents the examples also include messages regarding suggested pedagogy. We recommend limited use of examples within statements of grade-level learning expectations. Instead we urge authors to strive for clarity within the statement of the learning expectation. If additional information and/or guidance is needed for specific audiences (e.g., teachers or parents), we suggest that a supplement (or companion document) be developed for this particular purpose.
- *Involve people with a broad spectrum of expertise*. Many different constituent groups have valuable knowledge and expertise to contribute to the development of mathematics curriculum standards. These groups include: classroom teachers, mathematicians, curriculum supervisors, and researchers in the fields of mathematics education and cognitive psychology. We recommend that all voices be heard and taken into account in the development of appropriate and rigorous mathematics curriculum standards.
- *Collaborate to promote consensus.* Fifty states with 50 state standards documents increases the likelihood of large textbooks that treat many topics superficially. In order to increase the likelihood of focused curriculum materials, states will need to work together to create some level of consensus about important learning goals and expectations at each grade. This can be accomplished through state consortiums such as the New England Consortium that collaborates on common assessment specifications, through collaborative efforts sponsored by groups such as the National Council of Supervisors of Mathematics, the Association of State Supervisors of Mathematics, or the Council of Chief State School Officers. It can also be accomplished if states build their curriculum standards from a "core curriculum" offered by national groups such as the National Council of Teachers of Mathematics, the College Board, and/or Achieve, Inc. In fact, we recommend that a consortium of national groups collaborate to propose a national core curriculum that focuses on priority goals for each grade, K–8. In this way, states might still tailor their own curriculum goals and expectations around local needs while ensuring a much greater level of consistency across the states.

Clearly much work and effort has occurred at the state level for setting learning goals for mathematics. The state-level GLE documents present specific learning goals and also describe developmental trajectories for attaining these goals across the elementary years of schooling. For many states, GLEs represent a new level of state leadership for curriculum articulation. Although individual documents may provide increased clarity and coherence within their respective state, as a collection they highlight a consistent lack of national consensus regarding common learning goals in mathematics at particular grade levels.

## NOTE

1. A full report of the study is available.

## REFERENCES

Klein, D., Braams, B. J., Parker, T., Quirk, W., Schmid, W., & Wilson, W. S. (2005). *The state of state math standards*. Washington, DC: Thomas B. Fordham Foundation.

No Child Left Behind Act. (2001). *Public law no. 107-110*. Retrieved January 13, 2005, from http://www.ed.gov/policy/elsec/leg/esea02/index.html

CHAPTER 1

# STATE-LEVEL CURRICULUM STANDARDS

## Growth in Authority and Specificity

**Barbara J. Reys**

> One of the main functions of curriculum, as intended and as implemented, is to distribute the content of the curriculum throughout the days and years of schooling according to a coherent and reasoned set of goals. (McKnight et al., 1987)

Since the passage of the federal No Child Left Behind (NCLB) Act of 2001, state departments of education as well as local school districts have been scrambling to address the law's requirements. One major area of focus has been identification of student learning expectations in mathematics. These learning expectations, sometimes called curriculum standards, are referred to in recent state documents as grade-level learning expectations (GLEs). GLEs convey the specific mathematics content that students at particular grades are expected to learn (and teachers are expected to teach). Our intent in this document is to report on three studies of state GLEs conducted by teams of Center for the Study of Mathe-

*The Intended Mathematics Curriculum as Represented in State-Level Curriculum Standards: Consensus or Confusion?* 1–13

matics Curriculum (CSMC) researchers. To set the scene for and argue the importance of such studies, we begin by describing the national context in which these studies were conducted.

NCLB requires that states adopt "challenging academic content standards" in mathematics, reading/language arts, and science that (1) specify what children are expected to know and be able to do; (2) contain coherent and rigorous content; and (3) encourage the teaching of advanced skills (NCLB Act, 2001). Furthermore, states are required, beginning no later than school year 2005–2006, to measure the achievement of students against the state standards in grades 3 through 8. Therefore, state-level mathematics curriculum standards are foundational in that they provide guidelines of what will be assessed annually in mathematics. Increasingly, these documents are viewed as drivers of change at the district and classroom level. As Schmidt (2004) argues, "Curriculum is at the core of any educational system because it defines what schooling should accomplish; it specifies [in] what content areas no child is to be left behind" (p. 7).

In this report we briefly review the history of K–12 mathematics curriculum "standard-setting" in the United States. We also describe the present set of state-level K–8 mathematics curriculum standards documents in three content strands (Number and Operation, Algebra, and Reasoning) and examine how topics within these strands are sequenced and emphasized across the different state documents. For example, we examined the extent to which there is consensus regarding the grade placement of topics such as whole number and fraction computational fluency, linear equation solving, and attention to justification and reasoning throughout K–8 schooling. The Number and Operation strand was selected for analysis due to its historical importance and emphasis in the K–8 mathematics curriculum. The Algebra strand was selected because of current calls for greater emphasis in the elementary and middle grades. Finally, we were interested in understanding attention in the K–8 curriculum to processes such as reasoning. The research team intends to examine other strands of the mathematics curriculum as a follow-up to the study reported here.

## AUTHORITY FOR CURRICULUM STANDARD-SETTING

The question of what mathematics students in K–12 schools should learn has been a focal point of discussion for most of the history of American education (Cambridge Conference on School Mathematics, 1963; Conference Board of the Mathematical Sciences, 1975; National Committee on Mathematical Requirements, 1923; National Council of Teachers of Mathematics [NCTM], 1980; National Education Association [NEA], 1899; National Institute of Education [NIE], 1976). Documents detailing what students should learn in school in various areas of the curriculum have been developed by committees sponsored by professional organizations (e.g., NEA, AMS, MAA, and NCTM), the College Entrance Examination Board, and the NIE. However, the authority for making curricular decisions has historically resided at the local school district level with committees of teachers, community representatives, and parents working together to articulate the broad goals and desired outcomes of mathematics instruction within a district and then to select appropriate curriculum materials (textbooks) consistent with these goals (Long, 2003). In reality, school faculty and in some cases individual teachers determine what mathematics is emphasized in their school or classroom. In many cases, the textbook adopted for use in the school is a strong influence on the enacted curriculum. (Grouws & Smith, 2000; Weiss, Banilower, McMahon, & Smith, 2001).

In the 1980s, the role of state governments in defining mathematics curriculum standards grew. This coincided with the increasing role of states in measuring student learning for the purpose of public accountability. Prompted by national concerns such as those described in *A Nation at Risk* (National Commission on Excellence in Education, 1983) and international comparisons of student performance (e.g., the Second International Mathematics and Science Study), many states passed legislation raising standards for teacher certification, initiating new curriculum and assessment programs, and increasing graduation requirements (Long, 2003). In fact, over the past two decades, state standards, curriculum frameworks, and accountability measures have emerged as key state strategies for educational improvement (National Research Council, 2002, p. 39).

The focus on reforming mathematics curriculum was, in part, fueled by researchers who analyzed data from the Second International Mathematics Study and characterized the curriculum U.S. students study as an "underachieving curriculum." They reviewed the data for possible reasons for the relatively poor performance of U.S. students (e.g., class size, time spent on mathematics, and teacher preparation) and concluded:

> Of all the reasons offered, however, the culprit that seems to be central to the problems of school mathematics is the curriculum. It is the mathematics curriculum that shapes the textbooks that set the boundaries of instruction. It is the mathematics curriculum that distributes goals and content during the years of schooling. Something appears to be wrong with the way the content and goals are distributed in school mathematics in United States schools. (McKnight et al., 1987, p. 9)

The publication of *Curriculum and Evaluation Standards for School Mathematics* and *Principles and Standards for School Mathematics (PSSM)* by the

NCTM in 1989 and 2000 respectively had a strong influence on state-level mathematics curriculum development (Martin, Stein, & Ferrini-Mundy, 2002). However, state level mathematics curriculum guides produced in the early 1990s, like the first NCTM *Standards* document, were often general, specifying broad goals organized by grade bands rather than grade-by-grade descriptions of learning expectations. In the past 15 years, state documents have been reviewed and revised on regular cycles, tending toward greater specificity with each new version (Reys, Dingman, Sutter, & Teuscher, 2005). This trend toward greater specificity is driven, in part, by increased accountability in the form of state or federally mandated testing and, not coincidentally, by calls from teachers asking for more guidance in the mathematics on which to focus at particular grade levels. A review of past national, international, and state efforts underscores this movement to greater specificity and arguably greater authority.

## PREVIOUS ANALYSIS OF STATE-STANDARDS DOCUMENTS

Although documents published by NCTM in 1989 and 2000 influenced the content of state standards documents, states have generally worked independently of each other to create their own state mathematics curriculum framework. One notable exception is the set of New England states (Vermont, Rhode Island, New Hampshire, and Maine) that has worked together to produce near identical sets of GLEs.

The Council of Chief State School Officers (CCSSO) published reports in 1995 and in 1997 summarizing the content and quality of state curriculum standards documents as well as the processes used to develop the documents (Blank & Pechman, 1995; Blank, Langesen, Bush, Sordina, Pechman, & Goldstein, 1997). The 1997 report noted that learning expectations differed markedly by state but indicated that, in general, "state mathematics standards give a strong, consistent push for greater emphasis on higher level mathematics for all students, and less differentiation of curriculum for different groups of students" (Blank et al., 1997, p. vi).

Researchers involved in the Third International Mathematics and Science Study (TIMSS) (Schmidt, McKnight, Valverde, Houang, & Wiley, 1997) analyzed national standards documents in TIMMS countries and some U.S. state-level curriculum standards documents. They identified and described the organization of the official curriculum documents available at the time and noted the focus of each major section of the documents. Curriculum specialists in each country indicated the extent to which specified learning goals were included in their country's document. These researchers concluded that the U.S mathematics curriculum, as represented in state curriculum standards documents and textbooks, is a "mile wide and inch deep"—focusing on many topics with an excessive amount of repetition.

Given that publishing companies and textbook authors write materials for use in many states (arguably, all), the pressure to repeat content over several grades so that the materials fit differing state standards is great and may be a contributor to the growth in page count of textbook materials over the last century. Another influence on textbook authors and publishers is the need to comply with the demands of "state-adoption states." State-adoption states (22 states generally in the southern and western areas of the United States) evaluate curriculum materials for adoption at the state level. The process generally restricts schools and districts within the state from adopting materials (using state funds) not on the state-adopted list of text materials.

Two groups, one sponsored by the Fordham Foundation and another by the American Institutes for Research (AIR), have recently evaluated state mathematics standards documents. The Fordham Foundation review panel consisted of mathematicians who reviewed the official state mathematics curriculum documents using the following criteria: the standards' clarity, content, sound mathematical reasoning, and the absence of negative features. The panel found variation in the state documents and judged many to be of low quality. The report, *The State of State Math Standards* (Klein, Braams, Parker, Quirk, Schmid, & Wilson, 2005), notes nine "major problems" with state standards documents, including excessive emphasis on calculators, lack of attention to memorizing standard algorithms, too little attention to the coherent development of fractions in the late elementary and early middle grades, and excessive attention to data collection. The Fordham reviewers note, "few states offer standards that guide the development of problem solving in a useful way" (p. 11).

AIR researchers reviewed components of the educational system in the United States and Singapore, including curriculum frameworks, textbooks, assessments, and teacher preparation. Mathematics curriculum frameworks from seven states, representing about a third of the U.S. student population, were compared to the national curriculum of Singapore. The report, *What the United States Can Learn from Singapore's World-Class Mathematics System (and What Singapore Can Learn from the United States): An Exploratory Study* (Ginsburg, Leinwandet, Anstrom, & Pollock, 2005) notes that the United States state frameworks put more emphasis on "high-order thinking skills critical to competing in the 21st century" but that other elements of the educational system (textbooks, teacher training, assessments) are not well aligned or comparatively as strong as in Singapore.

NCLB has prompted a new wave of state-level curriculum articulation with specific attention to decisions about grade-specific learning expecta-

**Table 1.1. Publication Dates for Most Recent State-Level Mathematics Curriculum Documents (as of 2/1/06)**

| *Year* | *Number* | *States* |
|---|---|---|
| 2006 | 1 | MS |
| 2005 | 10 | AK, CA, CT, DC, HI, ID*, NV*, NY, ND, TX |
| 2004 | 15 | AR, DoDEA, GA, KY*, LA, ME, MD, MA, MI, MO, NH*, RI*, SD, VT, WA |
| 2003 | 8 | AL, AZ, KS, MN, NC, UT, WV, WY |
| 2002 | 5 | NJ, NM, OK, OR, VA |
| 2001 | 3 | OH, SC, TN |
| 2000 | 2 | IN, NE |
| Pre-2000 | 7 | CO, DE, FL, IL, MT, PA, WI |
| None | 1 | IA |
| | 52 | |

*Note:* In this table and the ensuing grade level expectations analyses, the District of Columbia and Department of Defense Educational Activity were treated as "states" due to the large student population served.
* Draft

tions in mathematics. In fact, about two-thirds of the states have published new documents in the past 3 years (see Table 1.1). In an effort to better understand the nature and role of the new mathematics curriculum standards documents, the CSMC initiated a series of activities and studies to organize, disseminate, and describe the current status of curriculum articulation in the United States. For example, CSMC developed an online database that provides links to each of the state mathematics curriculum standards documents (see http://mathcurriculumcenter.org/states.php). CSMC staff also conducted a survey of state mathematics curriculum specialists (employees of state departments of education) designed to describe the processes used to develop the most recent versions of state mathematics curriculum documents (see report at http://www.mathcurriculumcenter.org/reports_publications.php).

The third initiative, and the main focus of this report, is an analysis of state-level mathematics curriculum standards documents. The goal of the analysis was to describe the emphasis and grade-level placement of particular learning expectations as described in state curriculum documents. The focus was on two content strands (Number and Operation and Algebra) and one process strand (Reasoning). Because of the differing nature of the expectations in each of these content and process strands, the kinds of analyses undertaken and the methods used to identify the foci were different. Although we report the results of each of the three analyses in this document, it is important to note that the scope and methods of each of the three analyses were different. In *Number,* the focus was on a small set of identified topics judged to be central in the development of number across the grades—*fluency with basic number combinations (basic facts) for each operation; whole number computation; fraction concepts and computation; and computational estimation.* In *Algebra,* a comprehensive look was undertaken of the entire span of GLEs. The areas for more in-depth analysis arose from this initial comprehensive compiling and sorting of the GLEs. The areas that were selected out of this analysis were *patterns; functions; expressions, equations, and inequalities; operations; and properties*. In *Reasoning,* an initial examination of the state standards led to a classification of kinds of reasoning appearing in the GLEs—reasoning as conceptual understanding, reasoning used in problem solving, and reasoning for verification. From this preliminary analysis, reasoning *for verification* was selected for focus, primarily because GLEs related to verification were more easily identified than other dimensions of reasoning.

CSMC researchers sought to describe the amount of variation regarding the grade level(s) at which states called for particular learning goals/ expectations to be a primary focus of instruction. In other words, to what extent does the United States have consensus on the scope and sequence of particular topics within the strands studied? Two additional questions emerged as the state GLEs were examined, What is the nature of the grain size and what is the level of specificity of the GLEs in general?

## CURRENT STATE-LEVEL MATHEMATICS CURRICULUM STANDARDS DOCUMENTS

The current set of state-level mathematics standards documents, including those that articulate GLEs or secondary course-level learning expectations (CLE) can be found at http://mathcurriculumcenter.org/states.php. The compilation was based on a careful review of state department of education websites and a survey of state education officials. The search confirmed that multiple mathematics curriculum documents exist in many states, most developed over the past 15 years. The documents are referred to by different names such as *Content Standards, Curriculum Frameworks, Performance Standards/Indicators, Core Curriculum Standards,* or *Grade-Level Expectations* (see Appendix A for list of titles by state). They vary in their level of specificity, legal status (mandatory or voluntary), and purposes served. Some documents outline general learning goals across multiple subjects, while others describe specific curricular emphasis in subjects such as mathematics, and still others summarize the focus of state assessments.

**Table 1.2. Organization of K–8 Mathematics Grade-Level Learning Expectations (GLE) Documents by State and Grade Level (as of 2/1/06)**

| *Elementary/Middle School Learning Expectations* | | | |
|---|---|---|---|
| *GLE Documents (Grades K–8)* | *GLE Documents (Other Grades)* | *Grade-Band Documents* | *No GLE or Grade-Band Documents* |
| AL, AZ, AR, CT, DoDEA, DC, FL, GA, HI, ID, IN, KS, LA, MD, MI, MN, MS, MO, NV, NH, NM, NY, NC, ND, OH, OK, OR, RI, SC, SD, TN, TX, VT, VA, WA, WV, WY | AK (3–10)<br>CA (K–7)<br>ME (3–8)<br>NJ (3–8)<br>UT (K–7)<br>KY (4–8)[2] | CO[1] (K–4, 5–8, 9–12)<br>DE (K–3, 4–5, 6–8, 9–10)<br>IL[1] (early elem., late elem., middle/junior high, early HS, late HS)<br>MA (1–2, 3, 3–4, 5, 5–6, 7, 7–8)<br>MT (K–4, 5–8, 9–12)<br>NE (K–1, 2–4, 5–8, 9–12)<br>PA[1] (K–3, 4–5, 6–8, 9–10)<br>WI[1] (K–4, 5–8, 9–12) | IA |
| 37 | 6 | 8 | 1 |
| 51 | | | 1 |

*Note:* [1] CO, IL, PA, and WI have Assessment Frameworks dated 2003, 2004, or 2005 (CO [2003]: 3–10, IL [2004]: Grades 3–8, PA [2004]: Grades 3–8 and 11, WI [2005]: Grades 3–8 and 10)
[2] Since the KY document does not include Grade 3 GLEs, we did not include it in our analysis.

Although different types of curriculum documents have been produced and carry different authority within states, most states have documents that specify GLEs for mathematics and some specify high school mathematics course learning expectations. Table 1.2 provides a summary of the organization of GLE documents (as of February 1, 2006).

As noted in Table 1.2, 43 states have GLE documents that span K–8, K–7, 3–8, 4–8, or 3–10. Most of the state standards documents (26) were published since 2004 (see Table 1.1). States that did not have a GLE document at the time of this study included Colorado, Delaware, Illinois, Iowa, Massachusetts, Montana, Nebraska, Pennsylvania, and Wisconsin. Some of these states have subsequently developed documents that describe the mathematics to be assessed on state-level tests at each grade level.

The state GLE documents are organized by content strands similar to those outlined in *PSSM* (NCTM, 2000) including Number and Operation, Geometry, Measurement, Algebra, and Probability and Statistics. Some also include GLEs organized by process strands. The most common process strands are Problem Solving and Reasoning. The level of specific-

**Table 1.3. Number of GLEs per Grade in Content Strands of Selected State Standards Documents**

| | *K* | *1* | *2* | *3* | *4* | *5* | *6* | *7* | *8* | *Mean* |
|---|---|---|---|---|---|---|---|---|---|---|
| CA | 14 | 25 | 31 | 38 | 44 | 27 | 37 | 41 | — | 32.13 |
| FL | 67 | 79 | 84 | 88 | 89 | 77 | 78 | 89 | 93 | 82.67 |
| TN | 43 | 56 | 62 | 74 | 68 | 77 | 71 | 68 | 94 | 68.11 |
| NY | 28 | 56 | 45 | 52 | 56 | 67 | 64 | 64 | 48 | 53.33 |
| TX | 29 | 31 | 33 | 35 | 33 | 30 | 31 | 35 | 34 | 32.33 |
| MN | 16 | 18 | 26 | 26 | 25 | 26 | 30 | 27 | 32 | 25.11 |
| Mean | 32.8 | 44.2 | 46.8 | 52.2 | 52.5 | 50.7 | 51.8 | 54.0 | 60.2 | 49.5 |

ity or "grain size" of learning expectations varies across state documents. See Table 1.3 for a summary of the number of GLEs per grade in six state standards documents. These six states represent the range of number of learning expectations in the full set of documents. As noted, the mean number of GLEs per grade of this set of representative state standards documents is 49.5.

## AUTHORITY OF STATE-LEVEL CURRICULUM STANDARDS DOCUMENTS

The state documents noted in Appendix A represent the intended mathematics curriculum in each state. However, the degree of authority associated with each document differs. Several documents explicitly indicate that the content standards specified are required to be taught by mathematics teachers within the state. For example, Alabama's 2003 document includes "minimum and required" expectations:

> Content standards and related content included in bullets in this document are minimum and required (Code of Alabama, 1975, §16-35-4). They are fundamental and specific but not exhaustive. In developing local curriculum, school systems may include additional content standards to reflect local philosophies and add implementation guidelines, resources, and/or activities; which, by design, are not contained in this document. (p. iv)

Other documents do not explicitly state that the contents are mandated. Instead, they represent a recommended mathematics curriculum. These states encourage flexibility at the local district level to develop mathematics curriculum expectations based on state recommendations. For example, the following statement is found in the introduction to Florida's *Sunshine State Standards* (1996):

> These Grade Level Expectations are recommended, not mandated, by the state, but they will eventually become the basis for state assessments at each grade 3–10 in reading and mathematics. Districts that have already developed grade level expectations may continue to use them, but should ensure they correlate to the new state GLEs.... Districts, schools, teachers and students are encouraged to extend the content and achievement expected as they feel it is appropriate. (pp. 1–3)

Similarly, California's *Framework* (2005) indicates:

> The standards identify what all students in California public schools should know and be able to do at each grade level. Nevertheless, local flexibility is maintained with these standards. Topics may be introduced and taught at one or two grade levels before mastery is expected. Decisions about how best to teach the standards and in what order they should be taught are left to teachers, schools, and school districts. (chap. 2, p. 18)

Some of the state standards documents offer school districts the option to voluntarily implement the mathematics curriculum. For example, Maryland's 2004 document is titled, "Voluntary State Curriculum" and describes what students should know and be able to do at each grade level Pre-K through 8. Likewise, Missouri's Grade-Level Expectation document, published in 2004, is intended to

> represent the Department's latest effort to explicate the Show-Me Standards, in order to help local educators articulate precise learning outcomes for their students. District staff may use these expectations to strengthen alignment of their curricula to the Show-Me Standards, while Department staff will incorporate them into the soon-to-be-developed state model curriculum. (http://www.dese.state.mo.gov/divimprove/curriculum/GLEDocuments.html)

## RELATIONSHIP OF STATE-LEVEL CURRICULUM STANDARDS AND ASSESSMENTS

The state-level GLE documents are primarily meant to convey mathematical learning expectations for students. That is, they describe the mathematics that should be the focus of instruction and the goals for student learning in each grade. In addition to describing what mathematics should be learned, states are also mandated to assess student learning of the intended curriculum annually in grades 3–8 according to the NCLB mandate.

State-level GLE documents were reviewed to identify the relationship between the GLEs and the content to be assessed on the states' annual grade-level assessment. This relationship appears to differ by state. For example, in some cases, the GLE document conveys both the curriculum that is to be emphasized in instruction and the curriculum to be assessed locally and/or by the state. For example, the *South Carolina Mathematics Curriculum Standards* (2001) states:

> The Content Standards, which describe the mathematical knowledge, skills, and conceptual understandings expected of students, are meant to serve two purposes: they specify what should be taught and learned by all students in a grade, and they designate what should be assessed by South Carolina's teachers and by the State at each grade level. (p.1)

In other cases, the GLE document includes notes indicating a subset of learning expectations to be assessed. For example, the District of Columbia's *Standards for Teaching and Learning* (2002) includes both "content" and "performance" standards, the latter being an "assessment that gauge[s] the degree to which content standards have been attained" (p. 1). A subset of the content standards, called Essential Skills, "specify 'what students should know and be able to do'" (p. 1). The Essential Skills tested on the Stanford 9 Achievement Test, used as the state assessment, are noted with an asterisk within the GLE document.

As noted earlier, some states do not have a GLE document. Instead, these states specify at a broader grain size (e.g., grade bands) learning expectations for mathematics. For example, Colorado publishes learning goals organized by grade bands (K–4, 5–8, 9–12). They also publish the *CSAP Assessment Frameworks* (2003) that specify the mathematics to be assessed annually and warn that the assessment expectations are not to be taken as a specified curriculum. The introduction to each of Colorado's grade-level frameworks states: "While the assessment frameworks list the knowledge and skills that will be assessed by CSAP assessments at each grade, they are not new standards nor are they curriculum" (p. x).

Likewise, Pennsylvania's *Assessment Anchors* (2005) specifies the mathematics to be assessed on statewide assessments. Delaware's *State of Delaware Mathematics Curriculum Framework Content Standards* (1995) includes both grade-band content standards and "performance indicators" that specify the knowledge students should possess prior to assessments in grades K–10.

As noted earlier in this report, activity at the state level describing both the intended and the assessed curriculum is ongoing. For this report, we chose not to do a comprehensive analysis of the status of states' approaches to communicating learning expectations to be assessed. Instead, we have highlighted a sample of the various ways states are communicating with school districts regarding the assessed curriculum. Our focus is on the articulation of grade-level learning expectations that comprise the intended curriculum.

## METHOD OF ANALYSIS

Various techniques have been used to analyze or evaluate curriculum documents that specify student learning expectations (Ginsburg et al., 2005; Klein et al., 2005; Porter, 2004; Porter & Smithson, 2001). Unlike other researchers, our goal was to describe the content and grade placement of selected mathematics topics within the strands of Number and Operation, Algebra, and Reasoning across all state documents that specify grade-level expectations beginning no later than Grade 3 and continuing through Grade 7 or 8. Therefore, each team of researchers utilized or modified some techniques described by other researchers (e.g., "topic tracing" as used in the TIMSS analysis and described by Schmidt et al., 1997). Teams also developed methods particular to the mathematics content that was the focus. For example, within the analysis of the Algebra strand, the team first collected all GLEs pertaining to algebra from a sample of states; sorted these standards into sets that pertained to the same mathematical topic within algebra; and selected areas for systematic analysis from those that emerged as having enough emphasis in the documents to yield robust findings. These topics included patterns, functions, expressions, equations, inequalities, operations, and properties. The team developed coding schemes for sorting GLEs into these topic areas. The Number and Operation strand research team set out to describe the grade level that included the "culminating" learning expectation related to particular operations on numbers (e.g., fluency with fraction computation). To do this, the team collected and coded all learning expectations that referred to computation with fractions. Within the Reasoning strand, the team was interested in the extent to which curriculum documents included attention to reasoning across the various content strands. Therefore, a set of keywords related to reasoning (e.g., verify, justify, conjecture) were formulated and used to identify a set of learning expectations to be analyzed. The team further refined the area of focus *as reasoning for verification* when the first level of analysis showed that this was the primary area of focus for reasoning standards across the states.

A summary of the methodology used for each analysis is included in the chapters that follow. In general, three research teams, each consisting of two to six members, worked together to identify the mathematical topics for analysis and to develop strategies for collecting the relevant information from the state curriculum documents. In each case, learning expectations were the primary unit of analysis. That is, the set of learning expectations related to the mathematical topic were collected and coded utilizing a coding guide and form developed by the research team.

The initial task was to read through the set of state curriculum documents (see Appendix A for the set of state documents used in each analysis), compiling the set of GLEs that referred explicitly to particular topics identified in advance. These learning expectations were placed in a spreadsheet that included identification of state and grade level for each GLE. As the list of learning expectations were compiled, major themes were noted. Once the set of GLEs for a particular topic was finalized, the team developed a detailed coding scheme, utilizing the notes taken as the set of GLEs were compiled. For example, the team compiling GLEs with regard to whole number computation noted that in most states, the topic was introduced at a particular grade and grew in complexity over consecutive years of instruction with the culminating or final GLE noted at a particular grade. Some GLEs indicated the size of numbers (number of digits) to be computed while others specified particular strategies, algorithms, or tools. Trends such as these contributed to the development of a coding system to capture characteristics common to the learning expectations.

Next, the coding scheme was used to code all GLEs from a few states to judge the effectiveness of and refine the coding scheme. Once a coding system was finalized, the set of learning expectations that had been compiled were coded based on the criteria that had been discussed and agreed upon by members of the research team. In every case at least two members of the research team coded each set of GLEs, resolving discrepancies when needed. Once all GLEs had been coded, researchers then set out to summarize the key findings that emerged from the data.

In the chapters that follow, presentations of the data followed by summaries of the findings are shared.

CHAPTER 2

# ANALYSIS OF K–8 NUMBER AND OPERATION GRADE-LEVEL LEARNING EXPECTATIONS

**Barbara J. Reys, Shannon Dingman, Travis Olson, Angela Sutter, Dawn Teuscher, and Kathryn Chval**

This chapter provides a summary of findings related to analysis of the learning expectations within the Number and Operation strands of the state-level mathematics curriculum frameworks. The following questions guided the work:

> At what grades are learning expectations related to fluency with basic number combinations (basic facts), whole number computation, and fraction computation called for across state-level standards documents? To what extent does the grade placement vary across the state documents?
>
> What is the nature and extent of emphasis on concepts of fractions within the state curriculum standards documents? At what grades are these concepts introduced and emphasized and does this vary across state documents?

*The Intended Mathematics Curriculum as Represented in State-Level Curriculum Standards: Consensus or Confusion?* 15–57

> What is the nature and extent of emphasis on computational estimation within the state curriculum standards documents? At what grade levels do states call for emphasis on computational estimation and does this vary across the state documents?
>
> What messages related to the use of calculators/technology as computational tools are included within state standards documents? Do these messages vary by state and/or by grade level?

We did not attempt a comprehensive analysis of the Number and Operation strands within the state documents due to the scope of such an analysis. Rather, we chose a few topics within the Number strand that we felt were likely to be of interest to the field and then examined the treatment of these topics across the 42 state GLE documents available as of May 2005 (see Appendix A). The topics examined include:

- Fluency with basic number (single-digit) combinations;
- Multidigit whole number computation;
- Fraction concepts and computation;
- Computational estimation; and
- The role of calculators as computational tools.

Analysis of messages regarding calculators as computational tools was prompted by statements in a recent study sponsored by the Thomas B. Fordham Foundation entitled *The State of State Math Standards* (Klein et al., 2005). Authors of the report indicate a "major problem" with state standards:

> One of the most debilitating trends in current state math standards is their excessive emphasis on calculators. Most standards documents call upon students to use them starting in the elementary grades, often beginning with Kindergarten. (p. 10)

We sought to document and describe messages within the state documents regarding the role of calculators, particularly as computational tools, including the attention to calculators in the early elementary grades.

## ORGANIZATION OF LEARNING EXPECTATIONS

Concepts, skills, and applications of number and computation are major emphases of K–8 state standards documents. Indeed, across all state documents about a third of the total number of learning expectations at Grades K–8 are within number- and computation-related strands (see

Appendix A for a summary of the mean number of GLEs per grade by state and strand). The number-related GLEs are most prevalent in Grades K–6.

The state documents organize the number- and computation-related learning expectations within one or more strands, the most common being a strand called Number and Operation (a strand in 19 state GLE documents). The second most common strand title Number Sense is used in six state GLE documents. Other number strand labels include Numeration, Number Operations and Concepts and Arithmetic, Number and Operation Concepts. Five state documents include more than one number strand, one related to numerical concepts and the other related to computation and/or estimation. For example, the Alaska document includes two number-related strands: Numeration and Estimation and Computation.

Within the number- and computation-related strands, various subheadings are used to organize learning expectations, including number concepts, place value, number properties, number sense, number theory, computation, and estimation. A common organizational method, used in 10 state documents (Arizona, District of Columbia, Hawaii, Maryland, Missouri, New Mexico, Oregon, South Carolina, Tennessee, and West Virginia), mirrors that found in *PSSM* (NCTM, 2000):

- Understand numbers, ways of representing numbers, relationships among numbers and number systems.
- Understand meanings of operations and how they relate to one another.
- Compute fluently and make reasonable estimates.

## METHODOLOGY

The primary goal for the analysis of basic number combinations, whole number computation, and fraction computation was to identify the grade at which each topic is introduced, the developmental trajectory of the topic over subsequent grades, and the grade at which computational fluency is expected. Therefore, after compiling each set of GLEs from the appropriate state documents, individual GLEs were placed, one per row, into a spreadsheet along with an indication of the grade and source (state document). Columns were added to the spreadsheet to allow coding for major themes. For example, for the topic of basic number combinations, columns were added to the spreadsheet to allow coding of the initial GLE, intermediate GLEs, and the culminating GLE for each operation. The initial GLE referred to the first GLE related to basic number combinations at the earliest grade level. Subsequent GLEs referring to this topic were coded as "intermediate" GLEs. The GLE that indicated students were expected to know the full set of basic number combinations for an operation was coded as the "culminating" GLE for that operation. In addition, columns were included so that particular strategies or number size, if referenced in the GLE, could also be coded. For each topic, a unique coding sheet was developed to allow for nuances related to particular topics (further detail about coding related to each topic is included in later sections). Researchers used the spreadsheet to code a small set of GLEs to judge the effectiveness of and refine the coding scheme. The final coding system was used to code each individual GLE. Once all GLEs were coded, the research team then set out to summarize the key findings that emerged from the data.

The following sections include additional descriptions of the coding system related to specific topics as well as a summary of the major findings.

## FLUENCY WITH BASIC NUMBER COMBINATIONS (BASIC FACTS)

Fluency with basic number combinations represents a small, yet important emphasis of the Number strand, particularly in the primary grades. For this analysis, we examined all state GLE documents that spanned at least Grades K–6 (39 states). We did not include state documents that spanned, for example, Grades 3–8, since these documents would not include a full picture of the attention to basic number combinations.

Within the 39 state documents, 176 learning expectations that focus on the development of basic number combinations were identified. Most of the GLEs are within Grades 1–5. The set of GLEs utilize a variety of verbs to convey expectations, including *recall, know, memorize*, and *demonstrate automaticity*. The following examples illustrate the range of terms (and detail) used to convey goals related to addition and subtraction of basic number combinations.

> Demonstrate computational fluency for basic addition and subtraction facts with sums through 18 and differences with minuends through 18, using horizontal and vertical forms. (AL, gr. 2)
>
> Know the addition facts (sums to 20) and the corresponding subtraction facts and commit them to memory. (CA, gr. 1)
>
> State addition and subtraction facts. (AZ, gr. 2)

> Master addition and corresponding subtraction facts from 0 through 18. (DC, gr. 2)
>
> Demonstrate proficiency with addition and subtraction facts through 18.(ID, gr. 2)
>
> States and uses with efficiency and accuracy basic addition facts with sums from 0 to 20 and corresponding subtraction facts. (KS, gr. 2)

Based on multiple readings of the set of GLEs, a coding system was developed and applied. The system focused on the nature of the learning expectation. For example, does the GLE focus on use of models or various strategies for computing number combinations? Does it convey an expectation that students can accurately and efficiently recall a subset of number combinations? Does it convey an expectation that students are fluent with the full set of basic facts for a particular operation? Table 2.1 includes a summary of the codes for addition and multiplication as well as sample GLEs for each code. A similar coding system was used for GLEs related to subtraction and division of basic number combinations.

Seventy-nine of the 176 GLEs (45%) focused on understanding, modeling, or using strategies to compute the basic number combinations. In some cases, these learning expectations focused on a subset of number combinations (e.g., addition to 10). In others, use of strategies or manipulatives was specified to help build students' knowledge and understanding. Other learning expectations called for proficiency in recalling or constructing a particular set of combinations (or facts), such as "addition facts up to 9" or "multiplication combinations involving factors of 2 or 5." Still others, ones we called "culminating expectations," noted proficiency or fluency with a full set of combinations (e.g., multiplication facts to 9 × 9, 10 × 10, or 12 × 12). Generally, the culminating GLE signaled the end of attention to the topic and therefore was not mentioned or extended in later grades.

Figures 2.1 and 2.2 provide a summary of the grade at which fluency with basic number combinations for addition and multiplication is expected. The figures show, by state, the grade level where students are introduced to basic number combinations (initial GLE), where they are expected to advance their fluency by focusing on a small set of number combinations (intermediate GLEs), and where students are expected to be fluent with all basic number combinations (culminating GLE). We have not provided separate displays for subtraction and division because they are very similar to addition and multiplication. In fact, addition and subtraction are generally included within the same GLE, and the same is true for multiplication and division.

**Table 2.1. Coding Categories and Sample Statements of Learning Expectations Related to Basic Number Combinations for Addition and Multiplication (Basic Facts)**

| *Category* | *Sample Learning Expectation(s)* |
|---|---|
| Understand, model, or use addition strategies. (Initial Learning Expectation) | • Develop strategies for addition and subtraction basic facts such as counting on, counting back, making 10, doubles, and doubles plus one. (MD, gr. 1)<br>• Use strategies (e.g., count on, count back, doubles) for addition to at least sums to 12. (WA, gr. 1)<br>• Model addition through sums of 10 using manipulatives. (AZ, gr. K)<br>• Know the basic facts for addition and subtraction (0s, 1s, counting on and back 2s, doubles, doubles ± 1, then 10s facts, and related turn-around [commutative] pairs) and use them to solve real-life problems. (LA, gr. 1) |
| Know partial set of addition facts (e.g., facts to 9 or 10 or 12). (Intermediate Learning Expectation) | • Demonstrate computational fluency of basic addition and subtraction facts by identifying sums to 10 and differences with minuends of 10 or less. (AL, gr. 1)<br>• Apply with fluency sums to nine and related subtraction facts. (OR, gr. 1) |
| Know addition facts to 18 (or 20). (Culminating Learning Expectation) | • Know the addition facts (sums to 20) and the corresponding subtraction facts and commit them to memory. (CA, gr. 1)<br>• States and uses with efficiency and accuracy basic addition facts with sums from 0 to 20 and corresponding subtraction facts. (KS, gr. 2)<br>• Recall basic addition facts with sums up to 18 and the corresponding subtraction facts. (SC, gr. 1)<br>• Immediately recall and use addition, subtraction, and multiplication facts to 81. (NV, gr. 3) |
| Understand, model, or use strategies for computing multiplication number combinations (basic facts). (Initial Learning Expectation) | • Describe and compare strategies to solve problems involving multiplication and division (e.g., alternative algorithms, different strategies, decomposition, properties of multiplication). (WA, gr. 3)<br>• Models and multiplies numbers 0 to 5 using repeated addition. (MS, gr. 2)<br>• Use strategies (e.g., 6 × 8 is double 3 × 8) to become fluent with the multiplication pairs up to 10 × 10. (NM, gr. 3) |
| Know partial set of multiplication facts (2s or 5s or up to 5 × 5). (Intermediate Learning Expectation) | • States and uses with efficiency and accuracy the multiplication facts through the 5s and the multiplication facts of the 10s and corresponding division facts. (KS, gr. 3)<br>• Memorize basic multiplication facts 0–5 and the corresponding division facts. (WV, gr. 3) |
| Know multiplication facts (to 9 × 9 or 10 × 10 or 12 × 12). (Culminating Learning Expectation) | • Know the multiplication facts with automaticity to 10 × 10. (GA, gr. 3)<br>• Demonstrate automatic recall of basic multiplication and division facts through 100. (CO, gr. 4)<br>• Demonstrate mastery of the multiplication tables for numbers between 1 and 10 and of the corresponding division facts. (IN, gr. 4) |

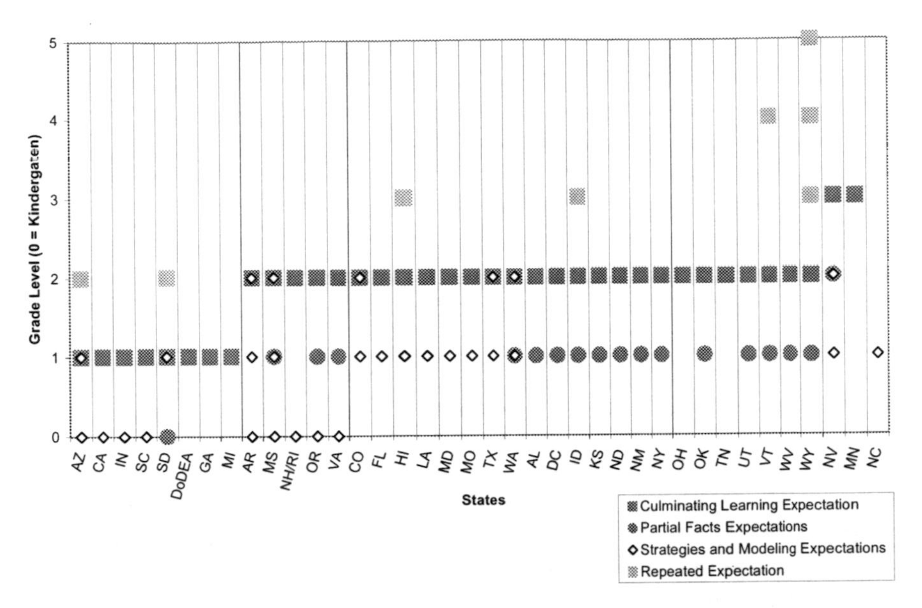


Figure 2.1. Progression to culminating learning expectations for addition of basic number combinations by state.

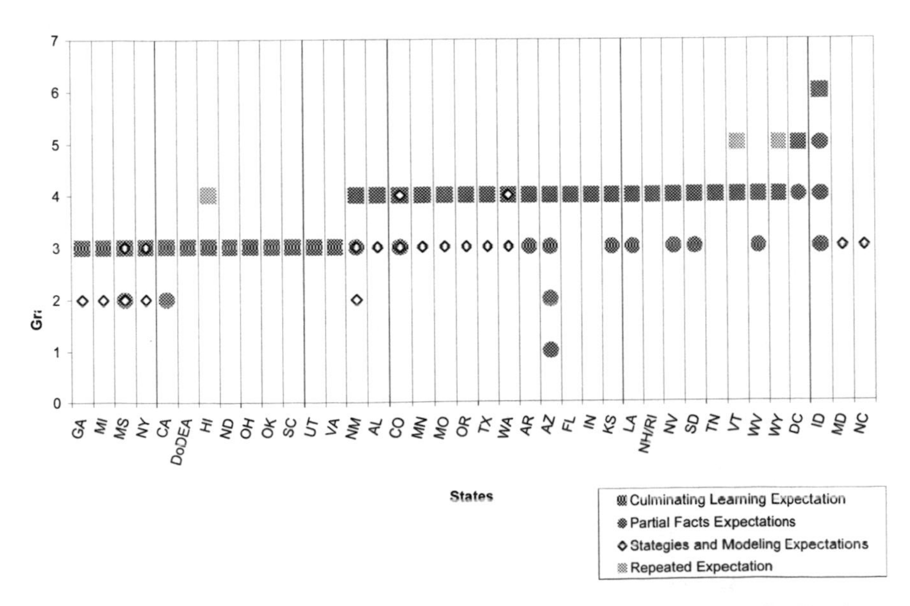


Figure 2.2. Progression to culminating learning expectations for multiplication of basic number combinations by state.

As noted in Figure 2.1, most state documents specify attention to basic addition combinations one year prior to when fluency is expected. For example, 20 state documents include GLEs introducing strategies or a partial set of addition basic number combinations in first grade and specify fluency one year later (Grade 2). Five state documents introduce the topic in kindergarten and expect fluency in Grade 1. With regard to multiplication number combinations (see Figure 2.2), five state documents introduce basic number combinations in Grade 2 and expect fluency at Grade 3, whereas 13 state documents introduce the topic at Grade 3 and expect fluency one year later (Grade 4).

Table 2.2 summarizes the grade placement of the culminating learning expectation for basic number combination fluency with each operation. That is, the grade at which state documents indicate fluency is expected. As noted, the most common grade placement for the culminating expectation for both addition and subtraction is Grade 2. The most common grade placement for multiplication and division combinations is Grade 4, although a significant number of states call for it at Grade 3.

The data confirm variation across the state documents with regard to when students are expected to begin to study and when they are expected to be fluent with basic number combinations. About two thirds of the state documents expect fluency with addition and subtraction number combinations by the end of Grade 2, and about a fourth call for this expectation at Grade 1 or 3. Likewise, states are fairly evenly split between Grades 3 and 4 with regard to the grade at which students are expected to be fluent with basic multiplication and division combinations. A discussion of the implications of this variation will be included later in this chapter.

**Table 2.2. Grade Placement of Learning Expectations Related to Fluency With Basic Number Combinations for Each Operation***

| *Operation* | *Grade* | *Number of States* | *Operation* | *Grade* | *Number of States* |
|---|---|---|---|---|---|
| Addition | 1 | 8 | Subtraction | 1 | 7 |
| | 2 | 28 | | 2 | 27 |
| | 3 | 2 | | 3 | 3 |
| | Not Specified | 1 | | Not Specified | 2 |
| Multiplication | 3 | 13 | Division | 3 | 6 |
| | 4 | 22 | | 4 | 20 |
| | 5 | 1 | | 5 | 3 |
| | 6 | 1 | | 6 | 1 |
| | Not Specified | 2 | | Not Specified | 9 |

*Note:* *For this analysis 39 of the 42 state documents were included (those that cover at least K-6).

## MULTIDIGIT WHOLE NUMBER COMPUTATION

Forty-two state documents were reviewed for this analysis, including 37 that span Grades K–8 and five that span a subset of Grades K–8 (e.g., Grades K–7 or Grades 3–8). Each state document was searched to identify all GLEs that mentioned computation with multidigit whole numbers for the operations of addition, subtraction, multiplication, and/or division. A total of 355 GLEs were compiled from the 42 state documents. The number of GLEs ranged from three in two state documents (Department of Defense Education Activity and Hawaii) to 19 in one state document (Utah). The following set of GLEs provides a sample of the progression of GLEs related to whole number computation in one state document (Indiana, 2000). As noted, the first GLE related to this topic in the Indiana state document is at Grade 2 and the final GLE is at Grade 4. The set also illustrates specifications with regard to number size (e.g., numbers less than 100) and type of strategy (e.g., use a standard algorithm) specified in some GLE statements. This specificity was not common to all state documents.

> Model addition of numbers less than 100 with objects and pictures. (IN, gr. 2)
>
> Add two whole numbers less than 100 with and without regrouping. (IN, gr. 2)
>
> Subtract two whole numbers less than 100 without regrouping. (IN, gr. 2)
>
> Add and subtract whole numbers up to 1,000 with or without regrouping, using relevant properties of the number system. (IN, gr. 3)
>
> Use a standard algorithm to multiply numbers up to 100 by numbers up to 10, using relevant properties of the number system. (IN, gr. 4)
>
> Use a standard algorithm to divide numbers up to 100 by numbers up to 10 without remainders, using relevant properties of the number system. (IN, gr. 4)

The set of GLEs was coded with attention to five categories: grade level, operation, size of number, specified strategies, and methods and tools of computation. The culminating learning expectation for each operation was also noted along with specifications regarding the size of numbers to be computed. Techniques noted in the GLE (e.g., explain, construct, use multiple strategies, or model/represent the computation of multidigit whole numbers) were coded as were the methods or tools for computation (e.g., mental computation, calculators, manipulatives, and/or paper and pencil). Finally, if a particular algorithm was noted in the GLE, this information was coded for analysis.

Twenty-five state documents include at least one GLE that focuses on explaining, constructing, using multiple strategies, or modeling multidigit whole number computation. Most of these are specified in the initial or intermediate GLEs. For example, 11 state documents include emphasis on explaining strategies for computing with multidigit numbers, most commonly at Grade 2. Examples include:

> Explain and perform addition and subtraction for two-digit numbers. (DoDEA, gr. 2)
>
> Apply and describe the strategy used to compute two-digit addition or subtraction problems. (MO, gr. 2)

Likewise, 12 state documents include GLEs that mention the use of multiple strategies to compute, most commonly at Grade 2. For example:

> Students are able to solve two-digit addition and subtraction problems written in horizontal and vertical formats using a variety of strategies (examples: doubles, near-doubles, one more/one less, etc.). (SD, gr. 2)

An expectation to model or represent multidigit computation (e.g., using concrete models or pictures) was included in 18 states and was most commonly found at Grade 3. For example:

> Using pictures, diagrams, numbers, or words, demonstrate addition and subtraction of whole numbers with two-digit numbers. (CO, gr. 3)

Specification of particular tools or methods of computation was noted in 24 state documents. For example:

> Selects and uses appropriate method(s) and/or tool(s) for computing with whole numbers from among mental computation, calculators, and paper/pencil according to the context and nature of the computation. (HI, gr. 4)

Few state documents specify the use of a particular algorithm for computing. However, seven of the 42 state documents include some GLEs that specify use of "standard" or "traditional" algorithms. For example:

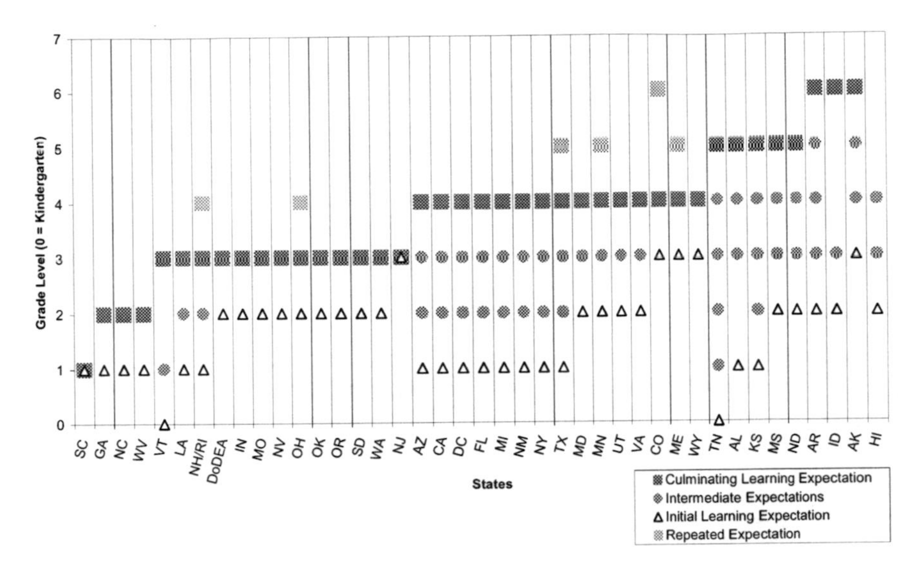


Figure 2.3. Progression of initial to culminating learning expectations for addition of multidigit whole numbers by state.

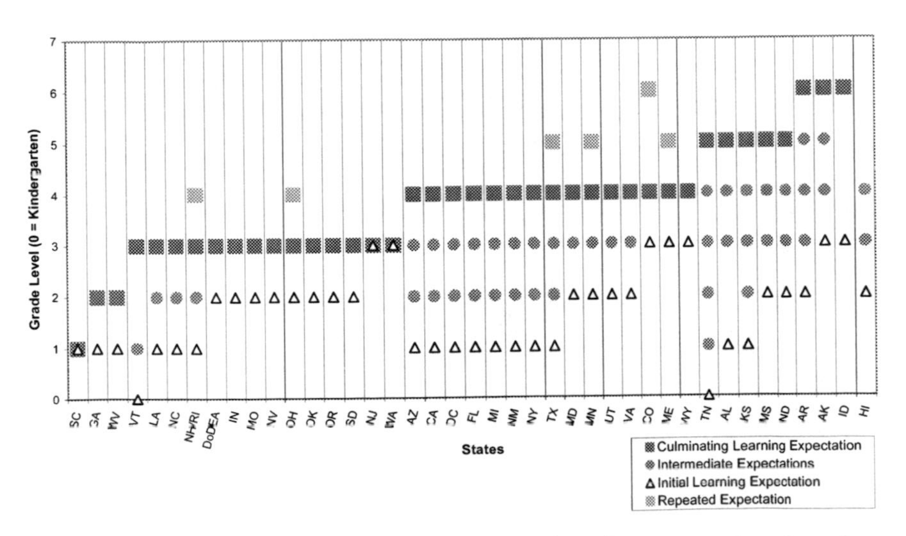


Figure 2.4. Progression of initial to culminating learning expectations for subtraction of multidigit whole numbers by state.

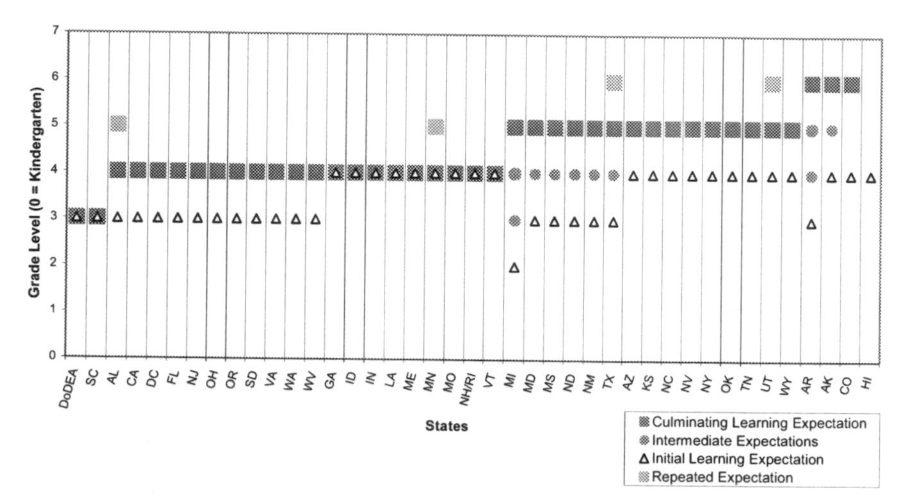


Figure 2.5. Progression of initial to culminating learning expectations for multiplication of multidigit whole numbers by state.

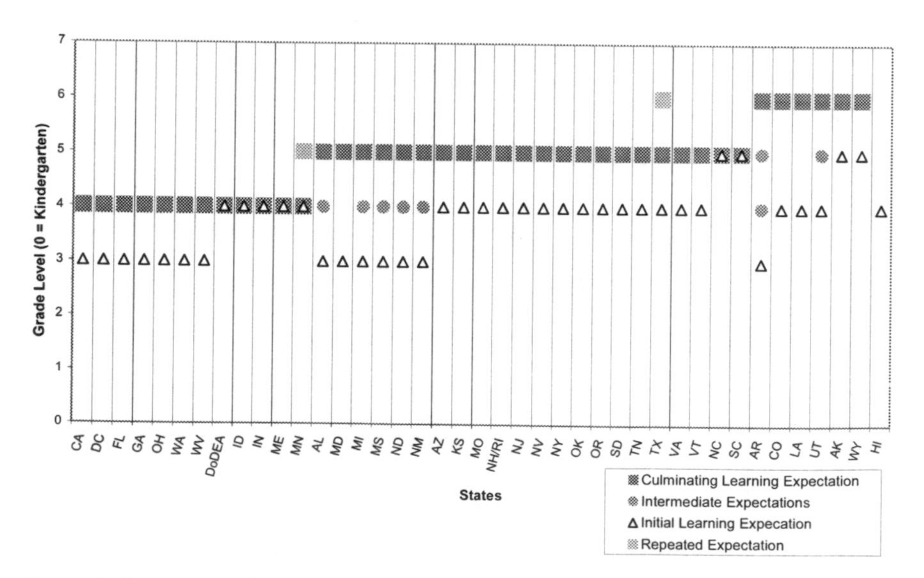


Figure 2.6. Progression of initial to culminating learning expectations for division of multidigit whole numbers by state.

**Table 2.3. Grade Placement of Culminating Learning Expectations Related to Fluency With Whole Number Computation for Each Operation**

| *Operation* | *Grade* | *Number of States* | *Operation* | *Grade* | *Number of States* |
|---|---|---|---|---|---|
| Addition | 1 | 1 | Subtraction | 1 | 1 |
| | 2 | 3 | | 2 | 2 |
| | 3 | 14 | | 3 | 15 |
| | 4 | 15 | | 4 | 15 |
| | 5 | 5 | | 5 | 5 |
| | 6 | 3 | | 6 | 3 |
| | Not Specified | 1 | | Not Specified | 1 |
| Multiplication | 3 | 2 | Division | 3 | 0 |
| | 4 | 21 | | 4 | 12 |
| | 5 | 15 | | 5 | 23 |
| | 6 | 3 | | 6 | 6 |
| | Not Specified | 1 | | Not Specified | 1 |

Demonstrate an understanding of, and the ability to use, standard algorithms for the addition and subtraction of multidigit numbers. (CA, gr. 4)

Figures 2.3–2.6 illustrate, by state, the grade at which students begin computing with multidigit numbers, all intermediate stages (e.g., with numbers of a certain size), and the grade at which students are expected to be fluent with the operation (i.e., culminating expectation). A few state documents repeat the culminating GLE in a later grade, and where this occurs, it is noted.

Figures 2.3–2.6 highlight the variation in grade placement of attention to whole number computation across the state documents. For example, in some states, students begin adding and subtracting multidigit numbers as early as kindergarten (labeled as Grade 0) or as late as Grade 3. The culminating GLE for addition and subtraction of multidigit whole numbers ranges from Grade 1 to Grade 6 across the states. The span between when the topic is introduced and when it culminates ranges from Grades 1–4, with two or three grades the most typical span. Multidigit multiplication and division are typically a focus at Grades 3 or 4 with fluency expected one year later (in Grades 4 or 5), although a few states expect fluency at Grade 6. Table 2.3 provides a summary of the grade placement of the culminating learning expectation for multidigit whole number computation for each operation.

## Specification of Size of Numbers Within Culminating Learning Expectation

Most state documents specify the grade at which students are expected to be fluent with whole number computation. Some also specify the size of numbers students should be fluent in computing. For example, the Indiana document states that by the end of Grade 3 students should "add and subtract two whole numbers up to three digits each with regrouping." On the other hand, the Ohio document indicates that students in Grade 3 should "add and subtract whole numbers with and without regrouping" but does not specify the size of numbers.

Nineteen of the 42 GLE documents (45%), while indicating the grade at which students should be fluent with whole number addition and subtraction, do not specify the size of numbers. On the other hand, 23 of the 42 (55%) state documents do specify size of numbers within the culminating learning expectation, ranging from two-digit numbers (New Hampshire, Rhode Island, South Carolina) to nine-digit numbers (Mississippi). The most commonly specified number size for addition and subtraction is three-digit numbers (nine state documents) and four-digit numbers (six state documents).

For multiplication and division of whole numbers, 11 of the 42 GLE documents (26%) do not specify the size of the numbers to be computed within the culminating learning expectation. Thirty-one of the 42 (74%) state documents do specify the size of the numbers to be computed. For multiplication, the size of numbers range from two-digit by one-digit numbers (Indiana) to three-digit by three-digit numbers (New York). Eight state documents specify that students should be able to multiply a two- or three-digit number by any size whole number (Alaska, Alabama, California, Michigan, North Carolina, North Dakota, Nevada, and Oklahoma). The most common specifications for multiplication are two-digit by two-digit numbers (12 state documents) and three-digit by two-digit numbers (seven state documents). For division, number size specifications range from two-digit dividend by one-digit divisor (Department of Defense Education Activity) to four-digit dividend by two-digit divisor (Kansas, Louisiana, Maryland, Michigan, South Carolina, Virginia, and Washington). The two most common specifications are division of any size dividend by a two-digit divisor (12 state documents) and division of a four-digit dividend by a two-digit divisor (seven state documents).

In summary, specification of whole number computation in the state standards documents varies with regard to (1) the grade at which instruction begins on multidigit whole number computation for each operation; (2) the grade at which students are expected to be fluent with multidigit whole number computation for each operation; and (3) specification of the size of numbers students are expected to compute with fluently as described in the culminating learning expectation.

We turn now to findings regarding the analysis of development of fraction concepts and computational fluency with fractions.

## FRACTION CONCEPTS AND COMPUTATION

The previous discussion focused on analysis of computation (basic number combinations and multidigit whole number computation for addition, subtraction, multiplication, and division) within the state GLE documents. It did not include a summary of the emphasis within the state documents related to number concepts (counting, place value, and so on) due largely to time constraints. However, for the analysis of fractions, we did examine GLEs that focused on conceptual development as well as computational facility.

Forty-two state documents were examined (the same set examined for whole number computation) for this analysis. These documents contained approximately 1,000 GLEs (mean of about 24 per state document) related to fractions spanning the full K–8 continuum. Common themes within the set include concept/meaning of a fraction; multiple representations of fractions; equivalence of fractions; conversions among fraction, decimal, and percent forms; and computation with fractions.

The level of detail and description regarding fractions within the state documents varies considerably. For example, the Washington document includes 47 fraction-related GLEs spanning Grades K–8, the highest number of any state GLE document. On the other hand, the Maine document includes eight fraction-related GLEs spanning Grades 3–8. Although GLEs related to fractions are generally spread across Grades K–8, the largest concentration are found in Grades 4, 5, and 6. In fact, 63% of the total set of fraction GLEs (635 of 1,009) are relatively evenly spread across Grades 4, 5, and 6, while 85% (853 of 1,009) are at Grades 3–7 (see Table 2.4).

The fraction GLEs were coded using the categories identified in Table 2.5. As noted, one distinction was whether the GLE focused on conceptual development or computational fluency. Some GLEs focused on both and were thus coded in multiple categories. GLEs focusing on fraction concepts were sorted into four major categories: (1) meaning of a fraction (a number that represents part of a set or whole, a representation of division, or as a point on the number line); (2) modeling or physically representing a fraction; (3) judging the size of a fraction (e.g., comparing, ordering, or using a number line, other model, or benchmarks to determine the size of a fraction); and (4) equivalence (simplifying fractions,

**Table 2.4. Number of Fraction GLEs and Percentage of Total by Grade**

| *Grade* | *Number of Fraction GLEs in all State Documents* | *Percent of Total Number of Fraction GLEs* |
|---|---|---|
| K | 7 | 0.7 |
| 1 | 45 | 4.5 |
| 2 | 72 | 7.1 |
| 3 | 117 | 11.6 |
| 4 | 195 | 19.3 |
| 5 | 232 | 23.0 |
| 6 | 208 | 20.6 |
| 7 | 101 | 10.0 |
| 8 | 32 | 3.2 |
| Total | 1009 | 100 |

**Table 2.5. Coding Categories and Example GLEs for Fraction Computation and Concepts**

| | *Coding Categories* | *Example Learning Expectation* |
|---|---|---|
| Fraction Computation | Add Fractions | Add or subtract fractions with like denominators (halves, thirds, fourths, eighths, and tenths) appropriate to grade level. (AZ, gr. 3) |
| | Subtract Fractions | Using concrete materials or pictures, add and subtract halves, thirds, and fourths. (CO, gr. 2) |
| | Multiply Fractions | Multiply and divide fractions and mixed numbers. (GA, gr. 6) |
| | Divide Fractions | Divide fractions and mixed numbers. (MD, gr. 7) |
| | Effect of Operations on Fractions | Describe the effects of addition and subtraction on fractions and decimals. (MO, gr. 6) |
| | Estimate with Fractions | Use strategies to estimate computations involving fractions and decimals. (NM, gr. 4) |
| Fraction Concepts | Meaning of Fraction | Use words, numerals, and physical models to show an understanding of fractions and their relationship to a whole. (DoDEA, gr. 2) |
| | Model/Represent Fractions | Identify alternative representations of fractions and decimals involving tenths, fourths, halves, and hundredths. (SD, gr. 5) |
| | Judge Size of Fractions | Compare and order whole numbers, fractions (rational numbers), and decimals and find their approximate locations on a number line. (UT, gr. 7) |
| | Equivalence | Apply factors and multiples to express fractions in lowest terms and identify fraction equivalents. (OR, gr. 6) |

converting from one fraction to an equivalent fraction, changing improper fractions to mixed numbers, and representing fractions as decimals and/or percents).

With respect to fraction computation, learning expectations were coded with regard to the operation (addition, subtraction, multiplication, and/or division), grade level, size of the denominator (if specified), and type of fraction (e.g., unit fractions, common fractions, proper fractions, mixed numbers, positive fractions, and fractions with common or uncommon denominators). Some terms such as "common fractions" are used within documents without definition or specification. In these cases, the term "common fractions" was defined by the research team to include fractions with denominators less than 20. Other GLEs coded under fraction computation included those that required students to estimate using fractions, model a specific arithmetic operation using fractions, and demonstrate the effect of an arithmetic operation on fractions.

## Findings Related to Fraction Concepts

This analysis focused on the set of learning expectations related to students' conceptualization or understanding of fractions. The data set includes 779 GLEs from 42 state documents and are summarized within three main categories:

- Understand fractions, including establishing the meaning of fractions, modeling or representing fractions, and communicating understanding of fractions.
- Equivalence across number formats, including simplifying fractions, changing improper fractions to mixed numbers, and converting fractions to decimals and percents.
- Judging the size of fractions, including comparing and ordering fractions.

*Understanding fractions*. A key element in the development of students' knowledge of fraction concepts is understanding the meaning of a fraction. Our analysis of this topic included examining GLEs focused on development of the meaning of a fraction, attention to modeling and/or representing fractions, and explaining/communicating understanding of fractions. These GLEs span the entire K–8 grade range, starting in the early grades as students learn what a fraction is and progressing through the middle grades as students continue to model situations involving fractions and explain their thinking. In this analysis, we examined GLEs that focused on:

- Fractions as a part of a whole or of a set of objects;
- Fractions as locations on a number line;
- Fractions as the division of whole numbers;
- Fractions as the remainder of a division problem;
- Fractions as the definition of rational numbers;
- Fraction terminology (i.e., numerator, denominator, halves, and so on); and
- Elementary fraction concepts (i.e., fractions equivalent to 1, comparing unit fractions, and so on).

Since it is common for the development of fraction concepts to begin as early as the primary grades (K–2), we limited this analysis to state documents that include grade-by-grade learning expectations from Grades K–7 or K–8 (39 state documents in all). That is, we excluded four states that do not include grade-by-grade GLEs for the primary grades. Thirty-seven of the 39 state documents contained GLEs pertaining to the development of the meaning of a fraction (Arizona and Idaho did not include GLEs for this topic). A total of 180 GLEs across the 37 states were identified for analysis, distributed as shown in Table 2.6.

Seventy-five percent of the GLEs in this set (135 of 180) focused on understanding the meaning of a fraction as a part of a whole or a part of a collection of objects. This idea dominated the set of the GLEs in the early grades. In fact, 16 of the 17 GLEs found in kindergarten, 30 of the 31 GLEs found in the first grade, and 35 of the 40 GLEs found in the second grade focused on understanding fractions as a part of a whole object or a set of objects. These GLEs were also found through Grade 6 with less frequency. The following GLEs are examples:

> Use physical models and drawings to represent commonly used fractions such as halves, thirds, and fourths in relation to the whole. (AR, gr. K)
>
> Share a whole by separating it into equal parts and use appropriate language to describe the parts such as three out of four equal parts. (TX, gr. 1)
>
> Recognize fractions as parts of a whole or parts of a group (up to 12 parts). (IN, gr. 2)
>
> Use a fraction to represent parts of a whole, division, or a ratio. (ND, gr. 6)

**Table 2.6. Distribution of GLEs Related to Understanding the Meaning of a Fraction**

| *Grade* | *K* | *1* | *2* | *3* | *4* | *5* | *6* | *Total* |
|---|---|---|---|---|---|---|---|---|
| Number of GLEs | 17 | 31 | 40 | 35 | 28 | 22 | 7 | 180 |

Twelve percent (22) of the GLEs, generally found in Grades 2–4, focused on fraction terminology, defining and using terms such as numerator and denominator. The following are examples of GLEs that concerned fraction terminology:

> Identify and use correct names for numerators and denominators. (IN, gr. 3)
>
> Introduces fraction terms and concepts including "fraction," "whole," "all," "part," "some," and "none." (MS, gr. 2)
>
> Use mathematical terms when communicating about computations involving fractions; i.e., numerator and denominator. (ND, gr. 4)

Eighteen GLEs (10%) focused on fractions as the division of whole numbers. These GLEs were typically found in Grades 4 (seven GLEs) and 5 (seven GLEs). Sixteen GLEs (9%) discussed elementary concepts related to fractions. These GLEs were typically found in Grades 2–5 and covered topics such as what happens to the value of a fraction as the numerator or denominator is changed, what happens when the numerator is equal to the denominator, and how fractions can be decomposed into unit fractions. Twelve GLEs (7%) focus on understanding fractions as points on a number line, the remainder of a division problem, and as the definition of rational numbers.

The state documents vary regarding the number and grade placement of GLEs related to the meaning of a fraction. For example, 24 state documents include GLEs related to the meaning of a fraction as early as kindergarten or Grade 1. Fifteen state documents also include GLEs focused on the meaning of a fraction in Grades 5 and 6. The number of GLEs covering this topic varies greatly across state documents, ranging from one GLE in both Hawaii and Missouri to 16 GLEs across Grades 1–6 in the Utah document.

As students gather an understanding of the different meanings of fractions, they also learn how to model and represent fractions and communicate their understandings of fractions.

With regard to modeling or representing fractions, 272 GLEs across the 42 state documents were identified. Approximately 78% of these

GLEs (212 of 272) were located between Grades 2–5 (44 in Grade 2, 54 in Grade 3, 66 in Grade 4, and 48 in Grade 5). The number of GLEs per state that discussed this topic ranged from one GLE for Alabama to 15 for Kansas. A majority of these 272 GLEs required students to use manipulative devices, models, or pictures to represent a given fraction, identify equivalent fractions, convert between equivalent number formats (fractions, decimals, percents) or to compare different fractions. The following are examples of GLEs that illustrate various purposes for modeling or representing fractions:

> Make models that represent given fractions (halves and fourths). (AZ, gr. 2)
>
> Model equivalent fractions with concrete objects or pictorial representations. (AL, gr. 3)
>
> Using concrete materials (e.g., fraction strips), compare and order fractions with like denominators, such as halves, thirds, fourths, eighths, and tenths. (CO, gr. 3)
>
> Represent with models the connection between fractions, decimals, and percents and be able to convert from one representation to another (e.g., use 10 by 10 grids, base-10 blocks, limit fractions to halves, fourths, fifths, and tenths). (OK, gr. 5)

Another aspect of the GLEs examined was the requirement that students communicate their understanding of fraction concepts. For this topic, we analyzed GLEs that indicated that students were to explain or describe their thinking about fractions. A total of 127 GLEs were identified across 29 states. Thirteen state documents (Alabama, District of Columbia, Georgia, Idaho, Minnesota, Missouri, New York, North Carolina, Oregon, South Dakota, Tennessee, Utah, and West Virginia) did not specify any GLEs indicating that students should communicate their understanding of fraction concepts, while eight state documents (Colorado, Maine, Michigan, Mississippi, Nevada, Minnesota, Oklahoma, and Virginia) included one GLE on this topic. Four states (Kansas, Washington, New Hampshire, and Rhode Island) accounted for almost a third of the GLEs (51 of 127) for this topic, with Kansas' document containing 18 GLEs, Washington's document containing 13 GLEs, and the New Hampshire/Rhode Island joint document containing 10 GLEs. Regarding grade-level emphasis, 60% of the GLEs (76 of 127) in this set were located in the Grade 4–6 band (24 at Grade 4, 30 at Grade 5, and 22 at Grade 6). An additional 30% of the GLEs were found in the Grade 1–3 band. Examples of GLEs that focus on communicating understanding of fractions include:

> Identify in symbols and in words a model that is divided into equal fractional parts (halves). (AZ, gr. 1)
>
> Explain the relationship of commonly used fractions to their equivalent forms, and explain their relationship to a whole. (DoDEA, gr. 3)
>
> Explain or show how a fraction can be decomposed into smaller fractions. (WA, gr. 4)
>
> Construct, use, and explain procedures for performing addition and subtraction with fractions and decimals with pencil-and-paper, mental math, calculator. (NJ, gr. 5)
>
> Knows and explains numerical relationships between percents, decimals, and fractions between 0 and 1. (KS, gr. 6)

*Equivalence across number formats*. One of the common concepts emphasized in the development of fractions is the equivalence of numbers represented in different forms. Our analysis included a search for GLEs related to simplifying fractions, changing improper fractions to mixed numbers, and converting fractions to decimals and percents. Many state documents begin this work by focusing on a limited set of fractions—ones they call "common" or "simple" fractions, which generally refer to fractions where the denominator is limited in size (e.g., less than 10 or 20).

We identified the GLEs in each state document that conveyed an expectation of ability to determine equivalence across multiple forms of fractions, decimals, and percents. If more than one GLE at different grade levels indicated students were expected to work with all fractions, we selected the GLE found in the lowest grade level to signify when students were expected to be proficient with this topic. Table 2.7 summarizes the grade level at which students are expected to move proficiently between equivalent forms of fractions, decimals, or percents.

For fraction-to-fraction equivalence, GLEs stating that students should be able to simplify fractions and judge the equivalence of any two fractions were found across five grade levels, ranging from third grade (Indiana) to seventh grade (Maine, Minnesota, and Nevada). The following are examples of the GLEs from various state documents related to fraction-to-fraction equivalence:

> Given two fractions, express them as equivalent fractions with a common denominator, but not necessarily a least common denominator

**Table 2.7. Number of States That Expect Proficiency at Each Grade Level**

| *Grade* | *Fraction to Fraction Equivalence* | *Improper Fraction to Mixed Number Equivalence* | *Fraction to Decimal Equivalence* | *Fraction to Percent Equivalence* |
|---|---|---|---|---|
| 3 | 1 State | 1 State | | |
| 4 | 10 States | 8 States | 5 States | 1 State |
| 5 | 22 States | 8 States | 5 States | 7 States |
| 6 | 6 States | 6 States | 23 States | 21 States |
| 7 | 3 States | 1 State | 8 States | 11 States |
| 8 | | | | 1 State |
| None Specified | | 18 States | 1 State | 1 State |

(e.g., 1/2 = 4/8 and 3/4 = 6/8); use denominators less than 12 or factors of 100. (MI, gr. 5)

Show equivalent fractions (fractions with the same value, e.g., 1/2, 2/4, 3/6, etc.) using equal parts. (IN, gr. 3)

As seen in Table 2.7, the expectation for changing improper fractions to equivalent mixed numbers was not specified explicitly in 18 of the 42 documents. Instead, this expectation may have been assumed within the fraction-to-fraction GLEs. The following are examples of the GLEs that explicitly called for converting improper fractions to mixed numbers:

Understand and use mixed numbers and their equivalent fraction forms. (NC, gr. 3)

Rewrite mixed numbers and improper fractions from one form to the other. (UT, gr. 5)

Determine equivalent forms of fractions, mixed numbers, and improper fractions. (OR, gr. 6)

Of the 24 state documents that included GLEs for this topic, 14 documents included the expectation at the same grade level as the fraction-to-fraction GLEs. As with the fraction-to-fraction GLEs, these learning expectations focused on converting between improper fractions and mixed numbers spanned Grade 3 (North Carolina) to Grade 7 (Minnesota). Figure 2.7 provides a summary of each state's placement of GLEs related to converting between fraction to fraction and improper fraction to mixed number.

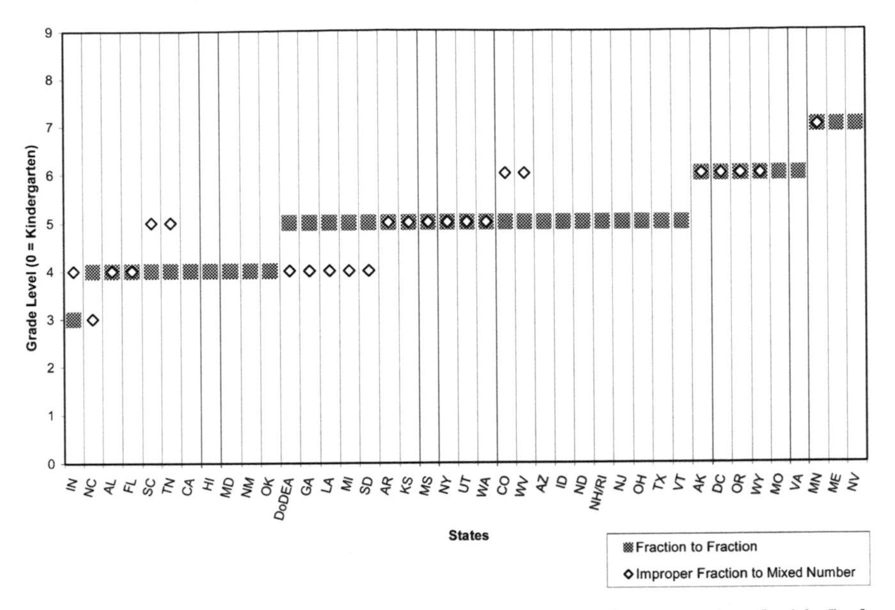


Figure 2.7. Grade level placement of learning expectations associated with finding equivalent fractions and converting improper fractions to mixed numbers, by state.

Figure 2.8 shows that 29 of the 41 state documents (the North Carolina document did not include GLEs for this topic) include learning expectations related to converting fractions to decimals and fractions to percents within the same grade level—one state document at Grade 4, two at Grade 5, 18 at Grade 6, and 8 at Grade 7. Example GLEs from these 29 states include:

> Convert between fractions, mixed numbers, decimals, and percents. (WV, gr. 6)
>
> Convert between any two representations of numbers (fractions, decimals, and percents) without the use of a calculator. (IN, gr. 6)
>
> Determine the equivalency among decimals, fractions, and percents (e.g., 49/100 = 0.49 = 49%). (AZ, gr. 4)

For the 12 states that separate attention to converting fractions to decimals and fractions to percents, 10 include these topics one grade level apart. Four state documents include conversion from fractions to decimals in Grade 4 and from fractions to percents in Grade 5, while three states

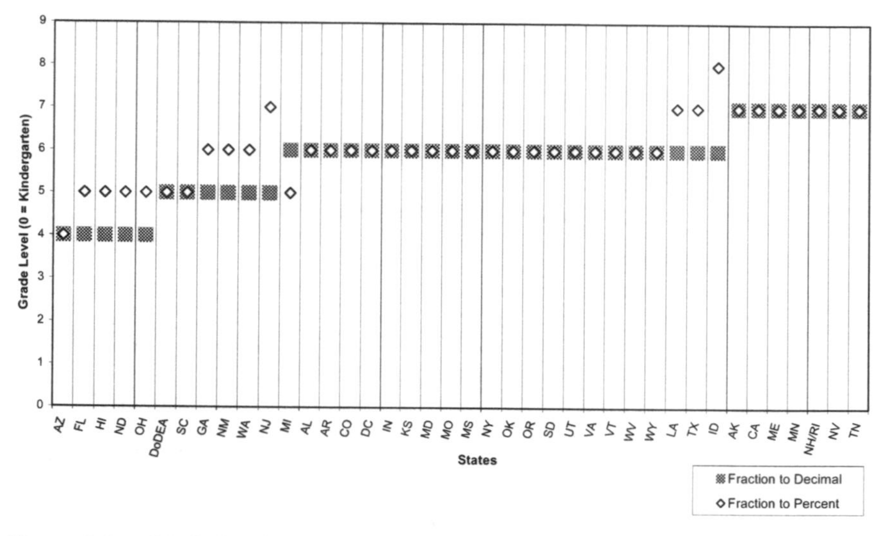


Figure 2.8. Grade-level placement for learning expectations dealing with converting fractions to decimals or percents.

used the same sequencing of topics for Grades 5 and 6 and another two states used this sequencing for Grades 6 and 7. Michigan was the only state found to hold the two topics in reverse sequencing, with the expectation of converting fractions to decimals in Grade 6 while the conversion from fractions to percents came in Grade 5. For the two other states (Idaho and New Jersey) that taught these topics separately, each held the topics two grade levels apart. See Figure 2.8 to examine the state's placement of these two topics.

*Judging the size of fractions.* One of the largest sets of fraction-related GLEs concerns judging the size of a fraction when compared with other rational numbers (i.e., whole numbers, other fractions, decimals, percents, mixed numbers). GLEs include language such as comparing, ordering, estimating, or using benchmarks or reference numbers. Some also included specification of particular representations (e.g., number line, models, or manipulatives). In all, 248 learning expectations across 42 state documents were examined pertaining to this topic.

Forty of the 42 state documents include attention to judging the size of a fraction by comparing it to another fraction. This GLE was found as early as Grade 1 and as late as Grade 6. Thirteen state documents initiate comparing fractions in Grade 3, while 11 state documents place the first GLE to compare fractions in Grade 4 and nine states start in Grade 2. The initial GLEs related to comparing fractions include varying emphasis such as:

**Table 2.8. Grade Where a Learning Expectation Related to Ordering Fractions First Appears**

| *Grade* | *K* | *1* | *2* | *3* | *4* | *5* | *6* | *7* | *8* |
|---|---|---|---|---|---|---|---|---|---|
| Number of States | — | 1 | 3 | 9 | 13 | 6 | 5 | 1 | — |

> Recognize, name, and compare unit fractions from 1/12 to 1/2. (CA, gr. 2)
>
> Compare fractions or mixed numbers with or without using the symbols. (MD, gr. 4)

Thirty-eight of the 42 state documents contained GLEs related to ordering fractions, in some cases with number lines or models. Four state documents (Georgia, Indiana, New Jersey, and the District of Columbia) do not include a GLE related to ordering fractions. Six other states have one GLE pertaining to ordering fractions. The Kansas document includes 12 GLEs spanning Grades 1 to 8 focused on ordering fractions. A GLE related to ordering fractions occurs for the first time at Grade 1 in the Kansas document and for the first time at Grade 7 in the South Dakota document. Table 2.8 summarizes the grade at which state documents include, for the first time, a learning expectation related to ordering fractions.

The variation in grade placement is, in part, explained by the particular focus of the learning expectation. For example, some state documents include an expectation that in the primary grades students order small or common fractions using models or manipulatives. For example,

> Compare and order fractions including halves, thirds, and fourths using a model (e.g., fraction circles, pictures, egg cartons, fraction strips). (OK, gr. 3)

Other state documents include more general statements regarding ordering fractions. For example,

> Convert, compare, and order fractions (mixed numbers and improper fractions), decimals, and percents and find their approximate locations on a number line. (AR, gr. 6)

Thirty-two of the 42 state documents include reference to number lines for judging the size of fractions. Of the 92 GLEs across the 32 states that mention the use of number lines in working with fractions, 29 (32%) are found in the fourth grade, 18 (20%) in the sixth grade, and 17 (18%) in the fifth grade. Several different uses for number lines were mentioned in the GLEs. For example, students were to use number lines in order to locate and label particular fractions:

> Use a number line to approximate and label halves, thirds, and fourths in relationship to whole units. (WA, gr. 4)
>
> Locate fractions, decimals, and mixed numerals on a number line. (MS, gr. 6)

Other GLEs referred to the number line as a way to compare and order fractions:

> Use a number line to simplify, compare, and order fractions and mixed numbers. (AL, gr. 4)

Twelve of the 42 state documents included GLEs that focus on estimating and using benchmarks (generally whole numbers or unit fractions) to either compute with fractions, locate fractions on a number line, or to determine the relationship (i.e., greater than, less than, or equal to) to another number. For example,

> Use models, benchmarks, and equivalence to add and subtract fractions with like denominators. (DoDEA, gr. 4)
>
> Use referent numbers in estimating answers to adding and subtracting fractions and mixed numbers (e.g., $2\ 1/4 + 3/8 < 3$ since both 1/4 and 3/8 are less than 1/2). (OR, gr. 6)

Some GLEs refer to the use of benchmarks in order to judge the size of a fraction and to determine whether it is greater than, less than, or equivalent to another fraction, decimal, or whole number. For example,

> Relate the size of fractions to the benchmark fractions 0, ¼, ½, ¾, and 1. (SC, gr. 5)
>
> Use concrete models or pictures to show whether a fraction is less than a half, more than a half, or equal to a half. (TN, gr. 2)

Seventeen of the 42 state documents provide GLEs that articulate when students should be able to use symbols (i.e., $<$ , $>$ , and $=$) to compare or explain the relationship between fractions. In all, 35 learning expectations were located in these 17 states that implied the use of symbols, with Louisiana and South Carolina including the greatest number (four each). Twelve of the 35 GLEs were found in the sixth grade, with 10 additional GLEs located in the fifth grade. Examples of GLEs that articulated the use of symbols when comparing and working with fractions include:

> Compare commonly used proper fractions and terminating decimals using the symbols: =, <, >. (CO, gr. 5)
>
> Use order symbols to compare two fractions, two decimals, or two percents. (SC, gr. 6)

The final method for judging the size of fractions concerns the use of models, concrete objects, and manipulatives. Twenty-nine of the 42 state documents contain GLEs that explicitly include the use of models or other tools for judging the size of a particular fraction. In total, 80 GLEs were noted specifying the use of some sort of object when determining the size of fractions. Many of these GLEs were found at the lower grades—30 of the 80 GLEs were located in the fourth grade, 19 in the third grade, and 10 in both second and fifth grades. The models and objects found in these GLEs include a varied set of materials. For example, models (concrete/physical, real-world, linear, area/region, set, fraction, ratio, graphic/visual), pictures or drawings, fraction strips, familiar objects, pattern blocks, diagrams, fraction circles, egg cartons, base 10 blocks, hundreds boards, Venn diagrams, and manipulatives were all mentioned in one or more state documents as materials to be used in working with fractions.

## Fraction Computation

The goal of this analysis was to describe when and by what trajectory states expect students to be proficient in computing with fractions (the term "fractions" is used here to refer to numbers represented with a numerator and denominator in the form of proper or improper fractions or mixed numbers). In addition, we identified the grade level of the culminating learning expectation related to each operation with fractions. The culminating learning expectation signified that students were expected to be fluent (or proficient) with the specified operation with all types of fractions (common and uncommon denominators). See below samples of the set of GLEs noted in our analysis as "culminating GLEs."

> Demonstrate computational fluency with addition, subtraction, multiplication, and division of decimals and fractions. (AL, gr. 6)

> Add and subtract common fractions and mixed numbers with unlike denominators. (GA, gr. 5)
>
> Compute and model all four operations with whole numbers, fractions (including mixed numerals), decimals, and percents applying order of operations and do straight computation with these numbers and operations. (ME, gr. 7)
>
> Divide fractions and mixed numbers. (MD, gr. 7)
>
> Add, subtract, multiply, and divide common fractions and mixed numbers as well as fractions where the common denominator equals one of the denominators. (MN, gr. 6)

*Addition and subtraction of fractions*. Figure 2.9 provides a summary of the progression of addition and subtraction computation with fractions for each state. As noted, state documents differ in their trajectory regarding computation with fractions. For example, both Arizona and Colorado include GLEs for teaching addition and subtraction of fractions in the first grade, and both states include a culminating learning expectation in Grade 6. The set of GLEs related to addition and subtraction in the Arizona document (six in all) follows:

> Demonstrate addition and subtraction of fractions with like denominators (halves) using models. (AZ, gr. 1) (designated as the Initial Learning Expectation)
>
> Demonstrate addition and subtraction of fractions with like denominators (halves and fourths) using models. (AZ, gr. 2)
>
> Add or subtract fractions with like denominators (halves, thirds, fourths, eighths, and tenths) appropriate to grade level. (AZ, gr. 3)
>
> Add or subtract fractions with like denominators, no regrouping. (AZ, gr. 4)
>
> Add or subtract proper fractions and mixed numbers with like denominators with regrouping. (AZ, gr. 5)
>
> Add or subtract proper fractions and mixed numbers with unlike denominators with regrouping. (AZ, gr. 6) (designated as the Culminating Learning Expectation)

On the other hand, the South Dakota and Missouri documents contain only one learning expectation related to addition and subtraction of fractions across all grade levels. This LE was coded as both the initial and culminating LE:

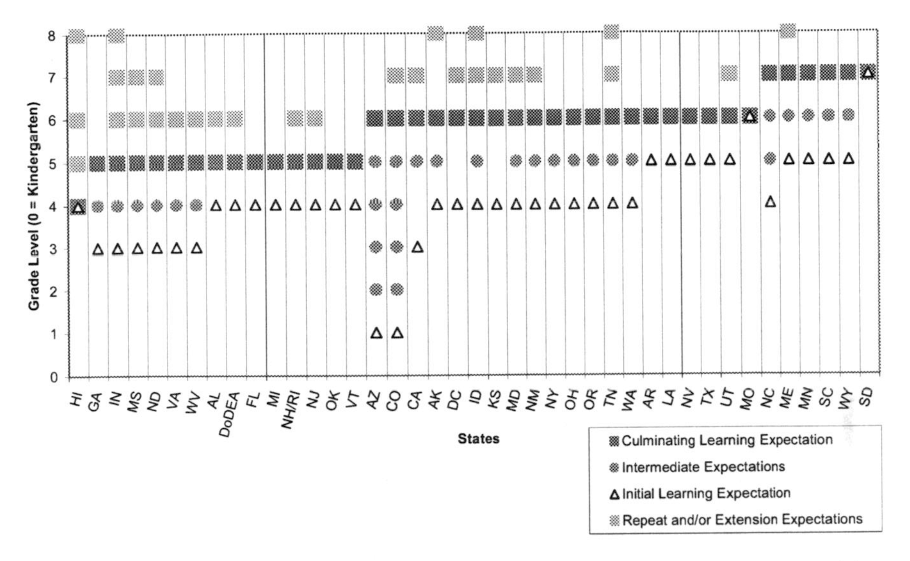


Figure 2.9. Progression of initial to culminating learning expectations for addition and subtraction of fractions, by state.

> Add, subtract, multiply, and divide integers and positive fractions. (SD, gr. 7)
>
> Add and subtract positive rational numbers. (MO, gr. 6)

*Multiplication and division of fractions.* Whereas many state documents spread the development of addition and subtraction with fractions across several grades, our analysis found that an expectation for fluency with multiplication and division of fractions often occurred within the same grade where it was introduced. Across the state documents, a similar range of grade levels was found for the culminating learning expectations for multiplication and division of fractions. However, unlike addition and subtraction, separate LEs were noted for each operation of multiplication and division.

For both multiplication and division, culminating learning expectations ranged from Grade 5 through Grade 8. For multiplication, 25 state documents specified the culminating learning expectation in Grade 6, while 13 state documents stated this in Grade 7. Two state documents (Florida and Hawaii) include the culminating learning expectation at Grade 5, while one state (Alaska) included the expectation for fluency with fraction multiplication at Grade 8. For division of fractions, 24 state documents specified the culminating learning expectation in Grade 6, with 13 state documents stating this in Grade 7. One state document

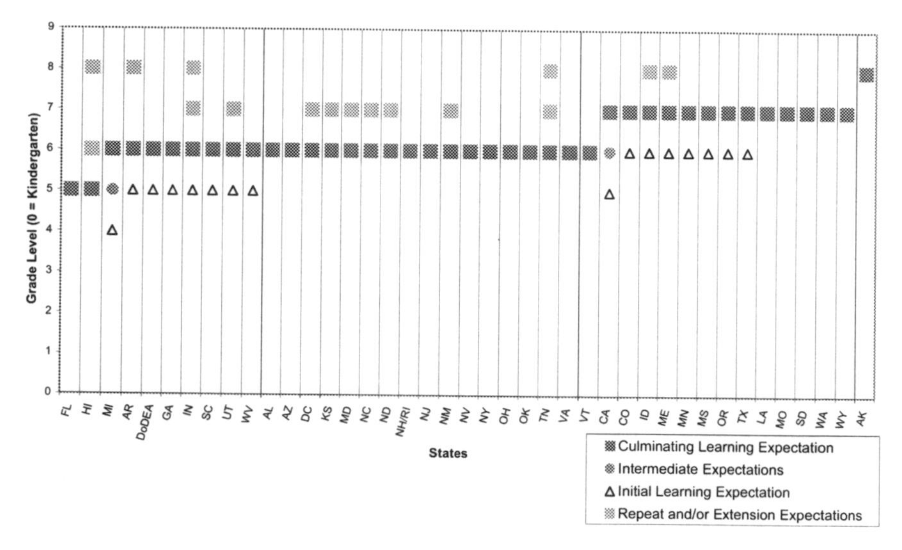


Figure 2.10. Progression of initial to culminating learning expectations for multiplication of fractions, by state.

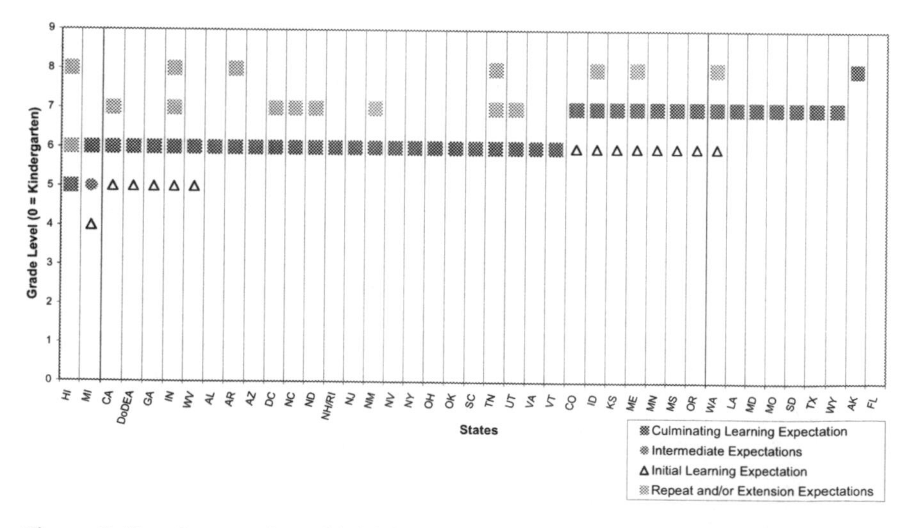


Figure 2.11. Progression of initial to culminating learning expectations for division of fractions, by state.

**Table 2.9. Summary of Grade Level When States Expect Students to Learn to Add, Subtract, Multiply, and Divide Fractions***

| *Grade* | *Addition of Fractions* | *Subtraction of Fractions* | *Multiplication of Fractions* | *Division of Fractions* |
|---|---|---|---|---|
| 4 | 1 State | 1 State | | |
| 5 | 15 States | 15 States | 2 States | 1 State |
| 6 | 20 States | 20 States | 26 States | 25 States |
| 7 | 6 States | 6 States | 13 States | 14 States |
| 8 | | | 1 State | 1 State |

*Note:* *For this analysis, we used the culminating learning expectation that indicated students were working with common and uncommon denominators when adding and subtracting fractions.

(Hawaii) held the culminating learning expectation in Grade 5, while another (Alaska) included division with fractions in Grade 8. No GLE related to division with fractions was found in the Florida document. See Figures 2.10 and 2.11 for a summary of the grade level for the culminating GLE for fraction multiplication and division by state.

In summary, the grade at which students are expected to be fluent with fraction computation varies across states. See Table 2.9 for a summary for each operation. Most state documents combine within the same GLE addition and subtraction of fractions. However, it is also common for state documents to specify GLEs for multiplication and division separately. The grade level where the culminating learning expectation for addition and subtraction of fractions was found ranged from Grade 4 (Hawaii) to Grade 7 (MAine, North Carolina, Minnesota, South Carolina, South Dekota, and Wyoming). That is, Hawaii's document conveys the expectation that students are fluent in adding and subtracting fractions at the end of Grade 4 whereas the Maine, North Carolina, Minnesota, South Carolina, South Dakota, and Wyoming documents include this expectation at Grade 7. The most common grade placement for this expectation is in Grade 5 (15 states) and Grade 6 (20 states).

## COMPUTATIONAL ESTIMATION

Forty-two state documents (those that include GLEs for Grades K–8 or another variation such as K–7, 3–8, or 3–10) were reviewed for this analysis. This summary describes the content emphasis and grade placement of computational estimation learning expectations in Grades K–8 and notes differences across state documents.

Learning expectations with references to estimation, finding approximate answers, and specific estimation strategies (e.g., "rounding") were

**Table 2.10. Grade-Level Learning Expectations Related to Computational Estimation in the New Mexico Mathematics Curriculum Document**

| *Grade* | *Learning Expectation* |
|---|---|
| 1 | Use and explain estimation strategies to determine the reasonableness of answers involving addition and subtraction. |
| 2 | Select and use a variety of appropriate strategies or methods to compute (e.g., objects, mental computation, estimation, paper and pencil). |
| 3 | Demonstrate reasonable estimation strategies for measurement, computation, and problem solving. |
| 4 | Use a variety of strategies (e.g., rounding and regrouping) to estimate the results of whole number computations and judge the reasonableness of the answers. |
| 4 | Use strategies to estimate computations involving fractions and decimals. |
| 5 | Use estimation strategies to verify the reasonableness of calculated results. |
| 5 | Explain how the estimation strategy impacts the result. |
| 5 | Recognize and explain the differences between exact and approximate values. |
| 6 | Use estimates to check reasonableness of results and make predictions in situations involving rational numbers. |
| 7 | Use estimation to check reasonableness of results, and use this information to make predictions in situations involving rational numbers, pi, and simple algebraic equations. |
| 7 | Convert fractions to decimals and percents and use these representations in estimations, computations, and applications. |
| 8 | Use a variety of computational methods to estimate quantities involving real numbers. |
| 8 | Approximate, mentally and with calculators, the value of irrational numbers as they arise from problem situations. |

identified for this analysis. The search focused on the number and operation and reasoning strands of the state documents, where 599 GLEs referencing computational estimation were identified. The Washington state document includes the greatest number of GLEs (54) related to computational estimation. The Vermont document includes a single GLE regarding estimating as a computational procedure for all grade levels (K–8):

> Estimates and evaluates the reasonableness of solutions appropriate to grade level. (VT, gr. K–8)

Only one of the 42 state documents (Maine) contained no GLEs related to computational estimation. The mean number of estimation-related GLEs is 14.3 per state document.

Learning expectations related to computational estimation from the New Mexico document are included in Table 2.10 to illustrate the emphasis and trajectory across grades in one state document.

The set of 599 computational estimation GLEs were coded according to the following categories: operation (addition, subtraction, etc.), type of number (whole number, fraction, decimal), estimation strategy (rounding, compatible number, etc.), and role of estimation (to predict approximate size of a calculation to be done, to judge reasonableness of exact calculation already completed). Also coded were GLEs related to rounding as a skill and general reference to benchmarks or referents. These later topics were summarized because of their prevalence in state documents and their relation to computational estimation.

The level of detail of GLEs related to computational estimation varies across state documents.

For example:

> The student determines reasonable answers to real-life situations, paper/pencil computations, or calculator results by using a variety of strategies (e.g., rounding to appropriate place value, multiplying by powers of 10, using front-end estimation) to estimate the results of whole number addition or subtraction computations to 10,000, or simple multiplication or division. (AK, gr. 4)

In contrast:

> Students use estimation strategies to solve problems. (WY, gr. 4)

Learning expectations related to estimation generally specify operations (addition, subtraction, multiplication, division) and types of numbers (whole, decimal, fractions, percents). As noted in Table. 2.11 most state GLE documents include learning expectations related to whole number, fraction and decimal estimation. Learning expectations related to estimation with percents are less common. The documents vary with regards to the grade at which computational estimation is first noted (see Table 2.12). For example, in six state documents, estimating sums and differences is noted as a learning expectation in Grade 1. On the other hand, other state documents include this learning expectation in Grades 4 or 5.

## Specification of Estimation Strategy

With regard to specification of computational estimation strategies, the state documents vary considerably. Most state documents (22 of the 42

**Table 2.11. Summary of Types of Numbers Specified in Computational Estimation GLEs by State**

| *Type of Number* | *Number of States* | *States* |
|---|---|---|
| Whole | 27 | AK, AL, AR, CA, CO, DC, DoDEA, FL, IN, KS, LA, MD, MI, MN, MO, MS, ND, NM, OH, OK, OR, SC, SD, TN, VA, WA, WV |
| Decimal | 28 | AK, AL, AR, CA, CO, DC, DoDEA, IN, KS, LA, MD, MN, MS, ND, NJ, NM, NV, NY, OK, OH, OR, SC, SD, TN, VA, WA, WV, WY |
| Fraction/Rational | 29 | AL, AR, CO, DC, DoDEA, FL, IN, KS, LA, MI, MN, MO, MS, NC, ND, NJ, NM, NV, NY, OH, OK, OR, SC, SD, TN, VA, WA, WV, WY |
| Percent | 13 | DC, DoDEA, LA, MS, ND, NJ, NM, NV, NY, OH, OK, OR, TN |

**Table 2.12. Grade at Which First GLE Related to Whole Number Estimation With a Specific Operation Occurs by Operation**

| *Operation* | *Grade* | *Number of States* | *States* |
|---|---|---|---|
| Addition | 1 | 6 | AR, CO, DC, NM, SD, WA |
| | 2 | 12 | FL, GA, IN, MD, MI, NC, ND, OH, OK, OR, VA, WV |
| | 3 | 8 | AK, AL, LA, MO, MS, NV, SC, TX |
| | 4 | 1 | CA |
| | 5 | 2 | NY, TN |
| Subtraction | 1 | 6 | AR, CO, DC, NM, SD, WA |
| | 2 | 9 | FL, GA, MD, NC, ND, OH, OR, VA, WV |
| | 3 | 11 | AK, AL, IN, LA, MI, MO, MS, NV, OK, SC, TX |
| | 4 | 1 | CA |
| | 5 | 2 | NY, TN |
| Multiplication | 3 | 7 | AR, CO, MS, NC, ND, NY, OK |
| | 4 | 11 | AK, AL, GA, LA, MD, MI, MO, SC, TX, VA, WA |
| | 5 | 2 | IN, NY |
| Division | 3 | 4 | AR, MS, NC, ND |
| | 4 | 11 | AK, AL, CO, LA, MD, NV, OK, SC, TX, VA, WA |
| | 5 | 4 | GA, IN, MO, NY |

analyzed) encourage use of "various" or "a variety" or "appropriate" strategies when estimating. In these instances, a specific estimating strategy is not specified. For example:

> Use strategies to estimate in problem-solving situations. (TN, gr. 3)

In other state documents, specific estimation strategies are identified. For example:

> Estimating sums and differences of whole numbers by using appropriate strategies such as rounding, front-end estimation, and compatible numbers. (AL, gr. 4)

The most common specified estimation strategy, noted in 21 state documents, is rounding. Other estimation strategies specified by at least seven states include compatible numbers, front-end estimation, and use of benchmarks to estimate.

A common GLE, appearing at multiple grade levels within 28 state documents, specified that students should "select and use appropriate methods and tools for computing including mental computation, estimation, paper and pencil, and calculators." Another common GLE referred to differentiating situations that call for estimates or exact answers. For example:

> Recognize when an estimate is more appropriate than an exact answer in a variety of problem situations. (TN, gr. 6)

Eighteen state documents include emphasis on ability to round numbers, generally beginning in Grades 2 or 3 (rounding whole numbers) and extending through Grade 5 (rounding decimal numbers). Thirty-five state documents include expectations related to determining the reasonableness of solutions. For example:

> Use appropriate estimation strategies in problem situations including evaluating the reasonableness of a solution. (WV, gr. 7)

While not common, seven state documents include GLEs that ask students to use estimation to predict a calculation. For example:

> Use estimation to predict computation results. (ID, gr. 3)

In summary, state documents address computational estimation in various ways, some emphasizing a particular operation, type of number, or estimation strategy, with others focusing on the role estimation plays in computation. As noted, the focus and specification of these GLEs varied across state documents.

## ROLE OF CALCULATOR AS COMPUTATIONAL TOOL

As noted earlier, authors of a recent Fordham Foundation report, *The State of State Math Standards* (Klein et al., 2005), indicate that attention to calculators is a "common problem" associated with state mathematics curriculum standards. They conclude:

> One of the most debilitating trends in current state math standards is their excessive emphasis on calculators. Most standards documents call upon students to use them starting in the elementary grades, often beginning with Kindergarten. (p. 10)

In order to document the attention to and specific messages about calculators in the 42 state-level GLE documents, we conducted an analysis of the statements of learning expectations within each document.

Of the 42 state GLE documents, 11 make no mention of the terms "calculator" or "technology" within the set of learning expectations. However, 4 of these 11 state documents include a statement (generally within the introductory material of the document) discussing the recommended role of calculators/technology within the mathematics classroom. For example, the following statement is included in the introduction of the Alabama (2003) document:

> Appropriate use of technology is essential for teaching and learning.... Technology enhances the mathematics curriculum in many ways, but is not intended to serve as a replacement for the teacher. The effective use of technology, however, does depend on the teacher. Teachers use technology in mathematics instruction to prepare students for an ever-changing world. The teacher makes instructional decisions about worthwhile investigative tasks that take advantage of technological aids. Technology influences the mathematics taught by providing exploratory opportunities and visual displays that would be tedious to generate by hand. Technology should be used to foster, rather than replace, the understanding of basic mathematical concepts. The use of appropriate technological tools provides support for all students to learn mathematics. Technology can be used by students and teachers to assess the understanding of meaningful mathematical concepts and to investigate more complex problems. (pp. 3, 6)

Within the remaining 31 state documents, each learning expectation with the word "calculator" or "technology" was compiled into a file with the specific state, grade level, and strand noted. A total of 451 GLEs utilizing one of the terms was identified.

The state documents vary with regard to the terminology used to convey messages related to calculators/technology. For example, one state document (Arkansas) uses "technology" exclusively within the GLEs to refer to various types of technology, including calculators. Eight state documents (Arizona, California, Hawaii, Idaho, Michigan, Oklahoma, Utah, and Virginia) include references to "calculator" but not to "technology." The remaining 22 state documents use both terms within the set of GLEs.

Of the 451 GLEs compiled from 31 state documents, 21 GLEs, or 5% (compiled from 7 state documents), stated that calculators were *not* to be used. For example:

> Calculate the area of parallelograms (squares and rectangles) without using calculators. (MS, gr. 6)
>
> Multiply and divide, without a calculator, numbers containing up to three digits by numbers containing up to two digits, such as 347 / 83 or 4.91 × 9.2. (MN, gr. 6)
>
> Convert between any two representations of numbers (fractions, decimals, and percents) without the use of a calculator. (IN, gr. 6)
>
> Demonstrate understanding of and proficiency with basic addition, subtraction, multiplication, and division facts without the use of a calculator. (CO, gr. 1, 2, 3, 4)

In addition, 34 of the 430 GLEs referred to use of software programs such as Excel or dynamic geometry software rather than to general calculators. For example:

> Identify and draw lines of symmetry in geometric shapes (by hand or using technology). (IN, gr. 3)
>
> The student recognizes and investigates attributes of circles, squares, rectangles, triangles, and ellipses using concrete objects, drawings, and/or appropriate technology. (KS, gr. K)

The remaining GLEs (396) formed the basis for our review. See Table 2.13 for a summary of the number of GLEs referencing calculators/technology by state. As noted, the Arkansas and Washington documents include the largest number (56 and 41, respectively) and six states (California, Hawaii, Idaho, North Dakota, Oklahoma, and Oregon) include only one or two GLEs referencing calculators across all grades. The mean number of GLEs referencing calculators in the 31 state documents is 12.8 per state (1.4 per grade), or a little less than 3% of the total number of GLEs per grade. If the Arkansas and Washington state documents are excluded, the mean drops from 12.8 to 10.3 calculator/technology GLEs per state document or a little over one per grade.

As shown in Table 2.13, the number of GLEs referring to calculators/technology is greater in the upper grades than the lower elementary grades. For example, the mean number of calculator/technology GLEs per grade at Grades K–2 is 0.59, at Grades 3–5 it is 1.40, and at Grades 6–8 it is 2.27. As noted, 11 of the 42 states represented in Table 2.13 have

**Table 2.13. Number of Calculator/Technology Learning Expectations per Grade by State**

| *State* | *K* | *Gr. 1* | *Gr. 2* | *Total Gr. K–2* | *Gr. 3* | *Gr. 4* | *Gr. 5* | *Total Gr. 3–5* | *Gr. 6* | *Gr. 7* | *Gr. 8* | *Total Gr. 6–8* | *Total Gr. K–8* | *Mean of State* |
|---|---|---|---|---|---|---|---|---|---|---|---|---|---|---|
| AL† | | | | | | | | | | | | | | |
| AK | | | | | 3 | 2 | 2 | 7 | 3 | 2 | 3 | 8 | 15 | 1.67 |
| AR | 2 | 2 | 2 | 6 | 4 | 5 | 4 | 13 | 6 | 13 | 18 | 37 | 56 | 6.22 |
| AZ | | | | | | | | | 1 | 1 | 1 | 3 | 3 | 0.33 |
| CA | | | | | | | | | 1 | | | 1 | 1 | 0.11 |
| CO | | 1 | 1 | 2 | 1 | 1 | 3 | 5 | 3 | 3 | 3 | 9 | 16 | 1.78 |
| DoDEA† | | | | | | | | | | | | | | |
| DC*† | | | | | | | | | | | | | | |
| FL | | 3 | 5 | 8 | 2 | 2 | 2 | 6 | 3 | 2 | 3 | 8 | 22 | 2.44 |
| GA | | 1 | 1 | 2 | 1 | 2 | 2 | 5 | 2 | 2 | 3 | 7 | 14 | 1.56 |
| HI | | | | | | 1 | 1 | 2 | | | | | 2 | 0.22 |
| ID | | | | | | | 1 | 1 | 1 | | | 1 | 2 | 0.22 |
| IN | | | | | | | | | 2 | 1 | 4 | 7 | 7 | 0.78 |
| KS | | 1 | 1 | 2 | 2 | 2 | 1 | 5 | 3 | 3 | 4 | 10 | 17 | 1.89 |
| LA | | | | | 1 | 1 | | 2 | | | 1 | 1 | 3 | 0.33 |
| MD† | | | | | | | | | | | | | | |
| ME† | | | | | | | | | | | | | | |
| MI | | | | | | | | | 1 | 1 | 1 | 3 | 3 | 0.33 |
| MN | | | | | | | | | 3 | 3 | 4 | 10 | 10 | 1.11 |
| MO† | | | | | | | | | | | | | | |
| MS | 1 | 1 | | 2 | 3 | 1 | 4 | 8 | 5 | | | 5 | 15 | 1.67 |
| NC | | | | | 1 | 3 | 1 | 5 | 1 | 1 | 1 | 3 | 8 | 0.89 |
| ND | | | | | 1 | | | 1 | | | 1 | 1 | 2 | 0.22 |
| NH/RI† | | | | | | | | | | | | | | |
| NM | | | 1 | 1 | | | | | | 1 | 4 | 5 | 6 | 0.67 |
| NJ | | | | | 4 | 4 | 3 | 11 | 4 | 5 | 5 | 14 | 25 | 2.78 |
| NV | 1 | 1 | 1 | 3 | 1 | 1 | 3 | 5 | 3 | 4 | 7 | 14 | 22 | 2.44 |
| NY | | | | | | | 3 | 3 | 1 | 3 | 1 | 5 | 8 | 0.89 |
| OK | | | | | | | | | | 2 | | 2 | 2 | 0.22 |
| OH | | | | | 1 | 1 | 1 | 3 | 1 | 2 | 2 | 5 | 8 | 0.89 |
| OR | | | | | | | | | | | 1 | 1 | 1 | 0.11 |
| SC | | | 3 | 3 | 2 | 2 | 3 | 7 | 1 | 1 | 1 | 3 | 13 | 1.44 |
| SD† | | | | | | | | | | | | | | |
| TN | | 1 | 1 | 2 | | 1 | 1 | 2 | 2 | 2 | 2 | 6 | 10 | 1.11 |
| TX | 3 | 3 | 3 | 9 | 3 | 2 | 4 | 9 | 1 | 2 | 4 | 7 | 25 | 2.78 |
| UT | | | 1 | 1 | 1 | 1 | 3 | 5 | 2 | | | 2 | 8 | 0.89 |
| VA | 1 | 4 | 2 | 7 | 1 | 7 | 5 | 13 | 3 | | | 3 | 23 | 2.56 |
| VT† | | | | | | | | | | | | | | |
| WA | | 2 | 5 | 7 | 4 | 5 | 3 | 12 | 6 | 9 | 7 | 22 | 41 | 4.56 |
| WV | | | | | | | | | | 3 | 5 | 8 | 8 | 0.89 |
| WY† | | | | | | | | | | | | | | |
| Total LE | 8 | 20 | 27 | 55 | 36 | 44 | 50 | 130 | 59 | 66 | 86 | 211 | 396 | 396 |
| Mean/Gr. level | 0.26 | 0.65 | 0.87 | 0.59 | 1.16 | 1.42 | 1.61 | 1.40 | 1.90 | 2.13 | 2.77 | 2.27 | 12.77 | 1.42 |

Note: * The DC document includes "technology integration" LEs, which span all content areas and include emphasis on learning about technology.
† Indicate state documents that do not reference calculators or technology within the statements of learning expectations.

**Table 2.14. Proportion of 396 GLEs That Reference Calculators/Technology by Content Strand**

| *Strand* | *Percent of GLEs* |
|---|---|
| Number and Operation | 56 |
| Algebra | 18 |
| Data Analysis and Probability | 10 |
| Geometry and Measurement | 4 |
| Other (process strands such as problem solving, communication, and reasoning) | 13 |

mathematics curriculum standards documents that contain no references to calculators within the set of GLEs. Another 18 of 42 states include 10 or fewer references to calculators within their document. Across all the documents, the largest concentration of references to calculators is in the middle grades. In fact, 211 of the 396 (53%) calculator-related GLEs identified are found at Grades 6, 7, or 8. Table 2.14 summarizes the number of GLEs referencing calculators/technology by content strand and indicates the Number strand contains the most references.

In addition to identifying the number of GLEs that reference calculators/technology, the analysis included a review of the intended role of the calculator within the GLEs. Six different categories were identified. See Table 2.15 for a list of categories, descriptions, and example GLEs.

Table 2.16 summarizes the number of GLEs assigned to each coded role. About one third of the 396 GLEs that reference calculators/technology focus on solving applied problems or equations and most of these are in the upper grades. A little over a fourth of the calculator/technology GLEs focus on using the tool to represent, model or graph mathematical ideas or data. Twenty percent of this set of GLEs reference calculators/technology as a tool for computing or estimating. That is, 79 of the 396 GLEs that include a reference to calculators/technology convey an intention that the tool will be used primarily for computation, and most of these (45 of 79) are at Grades 6–8.

These data suggest that calculators/technology are infrequently encouraged solely as a computational tool. The most prominent role for calculators/technology in Grades K–2 is for developing or demonstrating conceptual understanding, in Grades 3–5 for solving problems or equations, and in Grades 6–8 for representing mathematics.

Two other kinds of GLEs referred to calculators/technology. However, the focus was not on using calculators but rather on judgments made prior to or after the use of the tool. They include choosing an appropriate method of calculation and checking the reasonableness of calculated answers. Examples include:

**Table 2.15. Summary of Coding Scheme for Identifying Role of Calculator/Technology**

| *Category* | *Description* | *Examples* |
|---|---|---|
| Represent | Students use calculators/technology to represent mathematical quantities and ideas including different notations and graphs. They also connect physical models to mathematical language. | Represent and solve problem situations that can be modeled by and solved using concepts of absolute value, exponents, and square roots (for perfect squares) with and without appropriate technology. (AR, gr. 7)<br>Organizes, graphs, and analyzes a set of real-world data using appropriate technology. (FL, gr. 8) |
| Solve problems or equations | Students use calculators/technology to solve applied problems or equations. | Use calculator, manipulatives, or paper and pencil to solve addition or subtraction problems. (WA, gr. 2)<br>Use technology, including calculators, to solve problems and verify solutions. (NV, gr. 5–8) |
| Develop or demonstrate conceptual understanding | Students use calculators/technology to build conceptual knowledge of mathematical ideas and/or demonstrate understanding of these concepts. | Uses a calculator to explore addition, subtraction, and skip counting. (FL, gr. 1)<br>Understand the concept of the constant as the ratio of the circumference to the diameter of a circle. Develop and use the formulas for the circumference and area of a circle. Example: Measure the diameter and circumference of several circular objects. (Use string to find the circumference.) With a calculator, divide each circumference by its diameter. What do you notice about the results? (IN, gr. 6) |
| Analyze | Students use calculators/technology to compare, interpret, identify relationships, make predictions, interpret graphs, or make sense of data. | Read, interpret, select, construct, analyze, generate questions about, and draw inferences from displays of data. Calculators and computers used to record and process information. (NJ, gr. 6)<br>Uses technology, such as graphing calculators and computer spreadsheets, to analyze data and create graphs. (FL, gr. 7) |
| Compute or estimate | Students use calculators/technology to compute or estimate. | Use a variety of strategies to multiply three-digit by three-digit numbers (Note: Multiplication by anything greater than a three-digit multiplier/multiplicand should be done using technology.) (NY, gr. 5)<br>Generating sequences by using calculators to repeatedly apply a formula. (NJ, gr. 7–8) |
| Describe, explain, justify, or reason | Students use calculators/technology to help them describe strategies, explain reasoning, or justify mathematical thinking. | Use technology, including calculators, to investigate, define, and describe quantitative relationships such as patterns and functions. (NV, gr. 5–8)<br>The student communicates his or her mathematical thinking by representing mathematical problems numerically, graphically, and/or symbolically or using appropriate vocabulary, symbols, or technology to explain, justify, and defend strategies and solutions. (AK, gr. 7) |

**Table 2.16. Role of Calculator/Technology as Specified in Learning Expectations Within State-Level Curriculum Documents**

| *Role of Calculator/Technology* | *Grade Band* | *Number of States* | *Number of GLEs* | *Total GLEs** | *Percent of Total GLEs* |
|---|---|---|---|---|---|
| Solve problems or equations | K–2 | 6 | 16 | 130 | 33 |
| | 3–5 | 15 | 46 | | |
| | 6–8 | 21 | 68 | | |
| Represent | K–2 | 2 | 5 | 105 | 27 |
| | 3–5 | 11 | 17 | | |
| | 6–8 | 21 | 83 | | |
| Compute or estimate | K–2 | 2 | 3 | 79 | 20 |
| | 3–5 | 13 | 31 | | |
| | 6–8 | 15 | 45 | | |
| Develop or demonstrate conceptual understanding | K–2 | 6 | 19 | 64 | 16 |
| | 3–5 | 8 | 19 | | |
| | 6–8 | 11 | 26 | | |
| Describe, explain, justify, or reason | K–2 | 8 | 16 | 63 | 16 |
| | 3–5 | 8 | 18 | | |
| | 6–8 | 9 | 29 | | |
| Analyze | K–2 | 2 | 3 | 51 | 13 |
| | 3–5 | 5 | 7 | | |
| | 6–8 | 15 | 41 | | |
| Choose appropriate method of calculation | K–2 | 4 | 8 | 78 | 20 |
| | 3–5 | 15 | 36 | | |
| | 6–8 | 13 | 34 | | |
| Determine the reasonableness of a calculated answer | K–2 | 2 | 2 | 18 | 5 |
| | 3–5 | 4 | 9 | | |
| | 6–8 | 4 | 7 | | |

*Note:* * The number of GLEs does not sum to 396 because some are coded in multiple categories.

> Solve problems using the four operations with whole numbers, decimals, and fractions. Determine when it is appropriate to use estimation, mental math strategies, paper and pencil, or a calculator. (UT, gr. 5, 6)
>
> Use estimation as a tool for judging the reasonableness of calculator, mental, and paper-and-pencil computations. (SC, gr. 5)

Ninety-six of the 396 GLEs focus on checking the reasonableness of a calculated answer and/or choosing an appropriate method to calculate. Table 2.17 summarizes the number of instances by grade band. As noted, use of calculators for either of these roles is more frequent in the upper elementary or middle school years.

**Table 2.17. Summary of Learning Expectations Referring to Choosing Appropriate Methods of Calculation and Checking Reasonableness**

| *Tools* | *Grade Band* | *Number of States* | *Number of GLEs* | *Total GLEs* | *Percentage of Total GLEs* |
|---|---|---|---|---|---|
| Choose appropriate method of calculation | K–2 | 4 | 8 | 78 | 20 |
| | 3–5 | 15 | 36 | | |
| | 6–8 | 13 | 34 | | |
| Determine the reasonableness of a calculated answer | K–2 | 2 | 2 | 18 | 5 |
| | 3–5 | 4 | 9 | | |
| | 6–8 | 4 | 7 | | |

In summary, our analysis of the state mathematics curriculum standards documents does not support the conclusion offered in the Fordham Foundation report. We found only five state documents that include any references to calculators in the GLEs for kindergarten. In fact, about one fourth of the state documents include zero references to calculators in statements of GLEs at any grade level. Another 43% (18 of 42 documents) include 10 or fewer references to calculators across the set of elementary- and middle-grade GLEs.

A close examination of the GLEs that reference calculators reveals that the majority suggest calculators as tools for solving problems and/or representing data rather than as a replacement for facility with paper and pencil computation. It is also worth noting that references to calculators are concentrated at the middle grades. We found no indication that states advocate reliance on calculators at the expense of efficient mental or written procedures.

## SUMMARY

Findings from this analysis confirm that state mathematics curriculum documents vary along several dimensions including grain size (level of specificity of learning outcomes), language used to convey learning outcomes (understand, explore, memorize, and so on), and the grade placement of particular learning expectations. In particular, the grade at which particular topics are introduced, their trajectory of development, and the grade at which proficiency is expected differs dramatically across the state documents. For example, addition and subtraction with fractions is introduced and emphasized at different grade levels and the expectation for fluency varies by as much as three or four grade levels. The treatment of estimation also varies, with some states giving no attention to this topic and others integrating it throughout the development of exact computation. Messages regarding the use of calculators to develop numerical concepts and/or do computation also vary, with some states completely silent about the use of calculators/technology.

The state-level GLE documents lay out specific learning goals within the Number and Operation strands and also describe developmental trajectories for attaining these goals across the elementary years of schooling. For many states, grade-level learning expectations represent a new level of state leadership for curriculum articulation. In that sense, there is considerable effort to report in response to the "underachieving curriculum" described by Second International Mathematics and Science Study researchers and the "mile wide, inch deep" description of curriculum offered by TIMSS researchers. However, if these curriculum documents represent a new level of curriculum authority, then while there may be more coherence of the number and operation curriculum within states, at a national level mathematics curriculum goals remain a very mixed portrait.

CHAPTER 3

# ANALYSIS OF K–8 ALGEBRA GRADE-LEVEL LEARNING EXPECTATIONS

**Jill Newton, Gregory Larnell, and Glenda Lappan**

Over the past two decades middle school teachers and administrators have been under pressure to increase the number of students completing a course on algebra by the end of Grade 8. Two primary reasons are often cited by those who advocate for the study of algebra in the middle school: doing so offers students the opportunity to take advanced placement mathematics and statistics courses in high school; and studying algebra increases the cognitive demand of the middle school curriculum, which has been criticized as repetitive and superficial. Flanders (1987) conjectured that one reason for the failure of United States students to thrive in their first algebra course was the sudden increase in the amount of new mathematics curriculum material introduced as students transition from a typical middle school curriculum to a course on algebra. In fact, the curriculum of many high-performing countries (based on international comparisons of mathematics achievement such as the TIMSS) includes attention to algebraic concepts earlier than the United States (Mullis et

*The Intended Mathematics Curriculum as Represented in State-Level Curriculum Standards: Consensus or Confusion?* 59–87

al., 2001). Over the past decade, groups such as the National Council of Teachers of Mathematics (2000) have advocated increased attention throughout K–8 grades for consistent and coherent preparation for algebra.

Our main goal in analyzing the Algebra strand in the current state-level K–8 mathematics standards documents was to describe the nature and extent of emphasis on algebra and the variation across state documents. The following questions drove our analyses:

> What are the strands in which algebra expectations are found? What titles do states use to label the algebra grade-level learning expectations, K–8?
>
> What labels typically are used in state GLE documents to guide the development of the Algebra substrands, K–8?
>
> Within each identified substrand, what are major topics/concepts on which the states have agreement at particular grade levels?
>
> What are typical grade levels or grade bands over which a substrand has major emphasis?
>
> What is the scope of differences in grade placements for learning objectives related to algebra? What is typical?
>
> What is the range of levels of specificity of GLEs related to algebra?
>
> What constitutes the agreed-upon algebra learning expectations for K–8 across the state documents?

In this report, we first present a summary of the method of analysis then discuss the findings by algebra topics. We follow with a summary of findings using the research questions as a guide.

## ORGANIZATION OF ALGEBRA LEARNING EXPECTATIONS

The work involved in the analysis of the Algebra strand was completed in the following stages:

- Identifying major ideas/topics within the Algebra strand;
- Sorting/coding GLEs;
- Investigating the characteristics of the algebra GLEs within the substrands;

**Table 3.1. Most Common Words Used in Titles of K–8 Algebra Strand**

| *Primary Words in Titles* | *Number of Appearances* |
|---|---|
| Algebra/Algebraic | 45 |
| Functions | 21 |
| Patterns | 15 |
| Relations/Relationships | 7 |
| Reasoning/Thinking | 5 |

- Identifying variations in the algebra GLEs across state documents; and
- Summarizing the findings.

## Identifying Major Ideas/Topics Within the Algebra Strand

At the time of this analysis (Fall 2005), 49 of the 50 states (Iowa being the exception), along with District of Columbia, and Department of Defense Education Activity, had a document that specified mathematical expectations. Most of the documents are organized by content strand and have a strand specified for algebra-related content. The most common title for the strand is "Algebra," although many variations of this title are used across documents. Table 3.1 summarizes the primary words used in the algebra-related strand titles.

Twenty-eight K–8 standards documents further divide the Algebra strand into substrands. Seven of these documents (Department of Defense Education Activity, Missouri, New Mexico, Oregon, South Carolina, Utah, and West Virginia) use the Algebra substrands suggested in PSSM (2000) and several other states use modified versions of this set of substrands:

- Understand Patterns, Relations, and Functions;
- Represent and Analyze Mathematical Situations and Structure Using Algebraic Symbols;
- Use Mathematical Models to Represent and Understand Quantitative Relationships; and
- Analyze Change in Various Contexts.

Table 3.2 summarizes common words used in the Algebra substrand titles across all documents. We used the "common words" in the strand (Table 3.1), substrand titles (Table 3.2) and the *PSSM* (2000) Algebra sub-

**Table 3.2. Most Common Words Used in Titles of Algebra Substrands**

| *Word* | *Number of Appearances* |
|---|---|
| Patterns | 26 |
| Algebra/Algebraic | 22 |
| Function/Functional | 22 |
| Relations/Relationships | 19 |
| Models/Modeling | 14 |
| Equations/Equality | 12 |
| Symbol/Symbolic | 10 |
| Change | 9 |
| Representations | 7 |
| Expressions | 6 |
| Inequalities | 6 |

strands to guide the division of the algebra strand into substrands for the purpose of this analysis. These substrands include Patterns; Functions; Expressions, Equations, and Inequalities (EEI); Properties; and Relationships between Operations.

Forty states as well as District of Columbia and Department of Defense Education Activity were selected since at the time of our analysis their documents contained grade-level expectations (GLEs). The other states (Delaware, Kentucky, Maine, Massachusetts, Montana, Nebraska, New York, Pennsylvania, and Wisconsin) were not included because at the time of our analysis, they either provided grade-band standards (e.g., K–2, 3–5, and 6–8) or had grade-level expectations commencing later than third grade. Further use of the word *states*, will refer to the 42 GLE documents, which includes District of Columbia and Department of Defense Education Activity documents.

We searched the 42 documents (more than 3,000 GLEs) to identify GLEs that related to one or more of the five substrands. In a preliminary search we noted that some GLEs related to some of the substrands (e.g., Patterns) were located in the Number strand. Therefore, we searched each strand (not just the Algebra strand) to identify the full set of algebra-related GLEs.

Table 3.3 provides several examples of coding choices made during the analysis. The table includes examples of situations in which the GLE was coded in a single category and other situations in which the GLE included content that was coded into two categories.

Approximately 90% of the GLEs included in the Algebra strand were sorted into the categories noted in Table 3.3. Table 3.4 summarizes addi-

**Table 3.3. Examples of Algebra GLEs Coded by Substrand**

| *GLEs Sorted Into One Substrand* | | |
|---|---|---|
| *Substrand* | *Explanation* | *Example GLE* |
| Patterns | Sorting and classifying | Sort and classify objects by one attribute. (WV, gr. K) |
| | Using one term in a pattern to generate the next term | Using symbols and objects, identify, create, and extend a wide variety of patterns. (SC, gr. 1) |
| Functions | Relationship between two sets of data | Describe the relationship between two sets of related data such as ordered pairs in a table. (TX, gr. 4) |
| | Change (qualitative or quantitative) | Recognize and describe changes over time (e.g., temperature, height). (NJ, gr. 2) |
| Expressions, Equations, and Inequalities (EEI) | Translating from words to algebraic symbols | Translates problem-solving situations into expressions and equations using a variable for an unknown. (FL, gr. 4) |
| | Solving equations | Solve addition, subtraction, multiplication, and division equations with unknown numbers (e.g., $8x = 56$). (ND, gr. 3) |
| Properties | Properties with or without reference to algebra | Solve problems using the associative property of addition. (AL, gr. 2) |
| Relationships between Operations | Relationships between operations with or without reference to algebra | Understand various meanings of addition and subtraction, and the relationship between the two operations. (VA, gr. 2) |
| *GLEs Sorted Into Two Substrands* | | |
| *Substrands* | *Explanation* | *Example GLE* |
| Patterns and Functions | Movement from pattern (relationship between the terms of a pattern) to function (relating the position number to its corresponding term in the pattern) | Extend and recognize a linear pattern by its rules (e.g., the number of legs on a given number of horses may be calculated by counting by 4's, by multiplying the number of horses by 4, or through the use of tables). (CA, gr. 3) |
| | Actions applied to both patterns and functions | Represent and analyze patterns and functions using words, tables, and graphs.(DoDEA, gr. 4) |
| Patterns and EEI | Patterns requiring algebraic symbols | Describe linear, multiplicative, or changing growth relationships (e.g., 1, 3, 6, 10, 15, 21, ...) verbally and algebraically. (LA, gr. 7) |
| | Patterns and equations used to solve problems | Use algebraic expressions, patterns, and one-step equations and inequalities to solve problems. (NC, gr. 5) |
| Functions and EEI | One component involving function and one component involving EEI | Use tables, graphs, and equations to represent mathematics relationships and solve real-world equations. (CT, gr. 8) |
| | GLE indicates function, but the example indicates EEI | Understand that an equation such as $y = 3x + 5$ is a rule for finding a second number when a first number is given. Example: Use the formula $y = 3x + 5$ to find the value of $y$ when $x = 6$. (IN, gr. 4) |

**Table 3.4. Additional Categories for GLEs Within the Algebra Strand**

| *Category* | *Explanation* | *Example* |
|---|---|---|
| Modeling/ representation | GLE not specific about content being modeled | Use tools and strategies (e.g., manipulatives) to model problems. (ND, gr. K) |
| Number | Basic number skills were involved, and there was no indication that algebraic skills were needed | Write and solve number sentences from problem situations using addition and subtraction (e.g., You have three pencils and your friend has two pencils. You want to know how many pencils you have all together. Write a number sentence for this problem and use it to find the total number of pencils). (IN, gr. 1) |
| Symbol | GLE emphasized general mathematical symbols without a reference to algebra | Demonstrate understanding of the = sign as an equality symbol and explore inequalities and the symbol. (CT, gr. 3) |
| Unit analysis | Focus of the GLE was converting units of measurement | Convert one unit of measurement to another (e.g., from feet to miles, from centimeters to inches). (CA, gr. 6) |

tional categories established to sort the remaining algebra GLEs. Other than their mention here, these GLEs were not included in the analysis described in the remainder of this chapter.

*Number of GLEs.* Algebra GLEs represent approximately 17% of the total number of GLEs (across all strands) in the 42 state standards documents. Figure 3.1 shows the total number of algebra GLEs in the three largest substrands: Patterns; Functions; and EEI. The graph shows that the number is small in the early grades and steadily increases over Grades K–5, followed by a more dramatic increase over Grades 6–8.

Figure 3.2 shows the number of GLEs as sorted into the three major substrands. When the three areas of Patterns, Functions, and EEI are graphed on the same axes, we see the predominance of Pattern GLEs in Grades K–3 with a steady decline over Grades 4–8. The number of Function and EEI GLEs increase steadily over Grades K–4 with dramatic increases in EEI from Grades 5–8 and Functions from Grades 6–8. As shown, the largest set of GLEs in Grades 5–8 is related to Expressions, Equations, and Inequalities. The EEI strand represents what might be called "symbolic algebra," which suggests that algebra, particularly in the later grades, is focused on the development of symbolic algebra, or an equation-solving-driven algebra, more than on a function-based algebra.

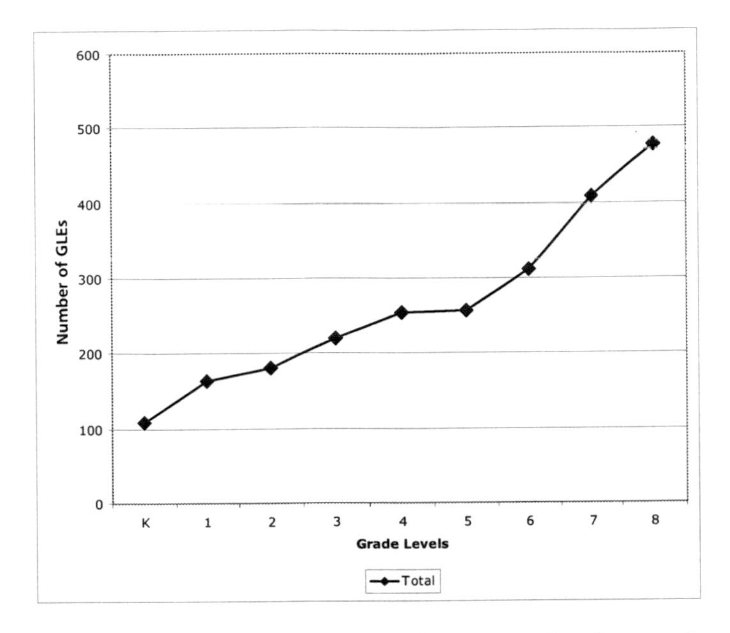


Figure 3.1. Total number of GLEs in patterns, functions, and EEI across grade levels.

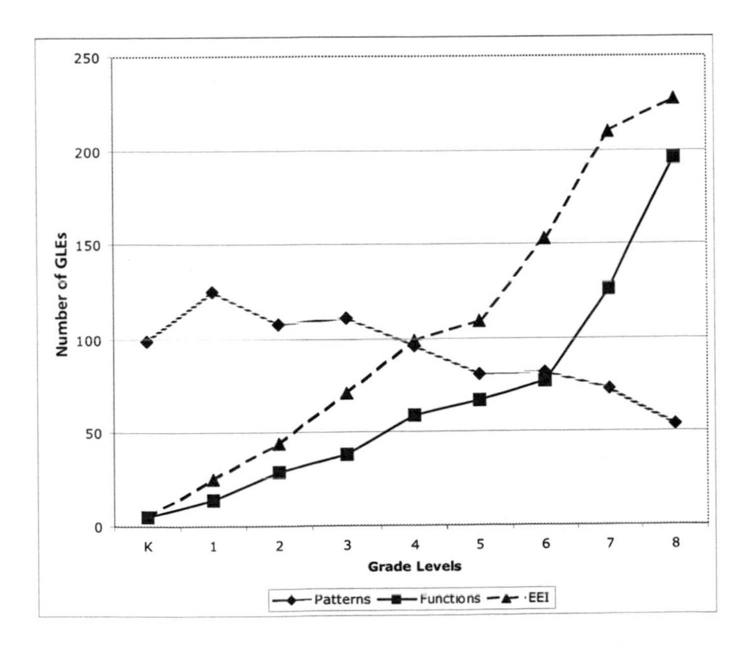


Figure 3.2. Total number of GLEs for three substrands (patterns, functions, and EEI across grade levels).

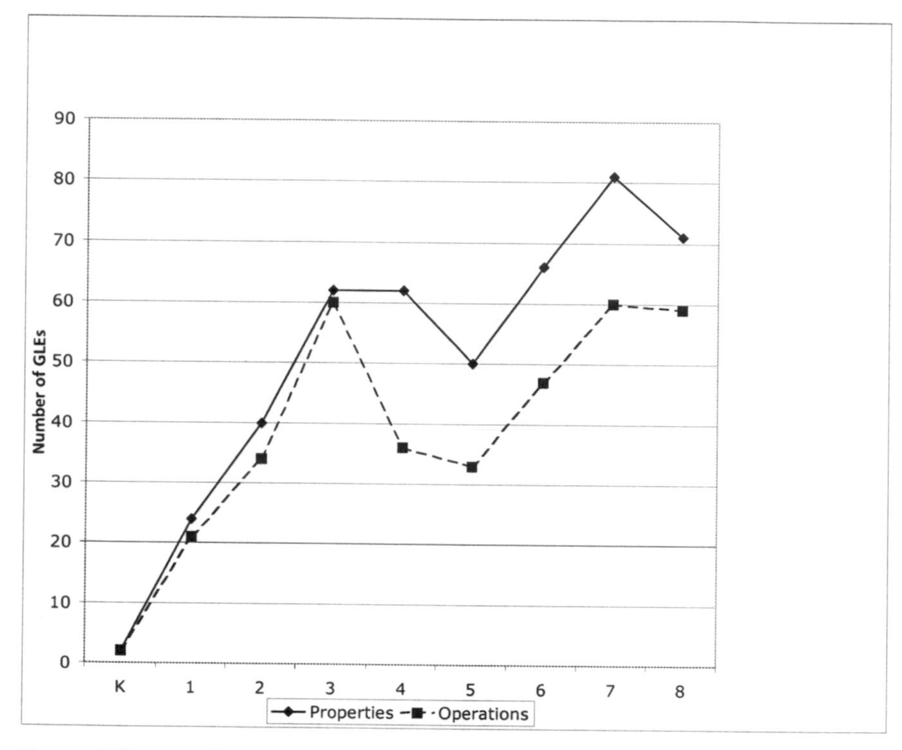


Figure 3.3. Number of GLEs related to properties and relationships between operations across K–8.

Figure 3.3 shows the number of GLEs in the other two substrands identified for this analysis across K–8, Properties and Relationships between Operations. The number of GLEs and trends of changes across grade levels are similar for each category. Since topics related to properties and operations are studied in the context of numeric manipulations and later in the context of algebraic manipulations, this graph and its increase–decrease–increase behavior is not unexpected.

Table 3.5 shows the information presented in Figures 3.1, 3.2, and 3.3 in numerical form. Both totals and percentages are given for each category and grade level (e.g., 113 GLEs representing 3.5% of the total algebra GLEs occurred in kindergarten).

In total, 3,193 algebra GLEs in the 42 state documents were identified for a mean of approximately 76 GLEs per state over Grades K–8. If they were evenly distributed, this would mean more than eight algebra-related GLEs at each grade level. Of course, the GLEs are not evenly distributed by states or by grade level within states. However, the

**Table 3.5. Total Number of Algebra GLEs by Grade Level**

| *Grade* | *Patterns* | *Functions* | *EEI* | *Properties* | *Operations* | **Total* | *Percent* |
|---|---|---|---|---|---|---|---|
| K | 99 | 5 | 5 | 2 | 2 | 113 | 3.5 |
| 1 | 125 | 14 | 25 | 24 | 21 | 209 | 6.5 |
| 2 | 108 | 29 | 44 | 40 | 34 | 255 | 8.0 |
| 3 | 111 | 38 | 71 | 62 | 60 | 342 | 10.7 |
| 4 | 96 | 59 | 99 | 62 | 36 | 352 | 11.0 |
| 5 | 81 | 67 | 109 | 50 | 33 | 340 | 10.6 |
| 6 | 82 | 77 | 153 | 66 | 47 | 425 | 13.3 |
| 7 | 73 | 126 | 210 | 81 | 60 | 550 | 17.2 |
| 8 | 54 | 196 | 227 | 71 | 59 | 607 | 19.0 |
| *Total | 829 | 611 | 943 | 458 | 352 | 3193 | 100 |
| Percent | 26.0 | 19.1 | 29.5 | 14.3 | 11.0 | 100 | |

Note: *As noted earlier, some GLEs were included in more than one strand and some Properties and Relationships between Operations GLEs were in the Number strand.

**Table 3.6. Most Common Verbs in the Algebra GLEs**

| *Patterns* | *Functions* | *EEI* | *Properties* | *Operations* | *Overall* |
|---|---|---|---|---|---|
| Extend (282) | Describe (145) | Solve (302) | Use (180) | Use (125) | Use (884) |
| Describe (229) | Use (125) | Use (278) | Apply (115) | Solve (60) | Describe (426) |
| Identify (185) | Identify (80) | Represent (127) | Demonstrate (26) | Apply (57) | Solve (443) |
| 52 Verb Families | 61 Verb Families | 73 Verb Families | 25 Verb Families | 34 Verb Families | 99 Verb Families |

mean provides an indication of the emphasis on algebra across the state documents. For comparison, the minimum number of algebra GLEs over the 42 documents is 30 (Wyoming) and the maximum number is 193 (Kansas) for an overall range of 163. The median number of expectations per state is 72.

*Common Format for GLEs.* Many of the algebra-related GLEs have a common format consisting of some combination of *verbs* (in italics), concepts (underlined), and **representations** (in boldface). For example,

> *Determine* the slope of a linear relationship *represented* **numerically or algebraically**. (MD, gr. 8)
>
> *Use* **models and words** *to describe, extend, and generalize* patterns and relationships. (DoDEA, gr. 4)

**Table 3.7. Examples of Verb Use Within Algebra GLEs**

| *Substrand* | *Verb* | *Example GLE* |
|---|---|---|
| Patterns | Use | *Use* multimedia resources to explore patterns, symmetry, and shape. (MS, gr. 1) |
| | Describe | *Describe* ways to get to the next element in simple repeating patterns. (MI, gr. 1) |
| | Solve | *Solve* problems from patterns involving positive rational numbers and integers using tables, graphs, and rules. (CO, gr. 7) |
| Functions | Use | *Use* mathematical models to show change in real context (e.g., create a graph showing how a candle's height changes over time after being lit). (ID, gr. 5) |
| | Describe | *Describes* relationships between two quantities that vary directly. (HI, gr. 4, 5) |
| | Solve | *Solve* simple problems involving a functional relationship between two quantities (e.g., find the total cost of multiple items given the cost per unit). (NM, gr. 3) |
| Expressions, Equations, and Inequalities (EEI) | Use | *Use* order of operations to evaluate and simplify algebraic expressions. (AL, gr. 8) |
| | Describe | *Describe* a real-world situation represented by an algebraic expression. (MD, gr. 8) |
| | Solve | *Solving* two-step linear equations of the form $ax \pm b = c$, where $a$, $b$, and $c$ are rational numbers, and a  0 or translating a story problem into an equation of similar form and solving it. (AK, gr. 8) |
| Properties | Use | *Use* commutative, associative, distributive, identity, and inverse properties to simplify and perform operations. (OH, gr. 5) |
| | Describe | *Describe* the commutative properties of addition and multiplication with words or symbols. (TN, gr. 3) |
| | Solve | *Solve* problems using the associative property of addition. (AL, gr. 2) |
| Relationships between Operations | Use | *Use* the inverse relationship between multiplication and division to compute and check results. (CA, gr. 3) |
| | Describe | *Describe* models of sharing equally (division) as repeated subtractions and arrays. (SC, gr. 2) |
| | Solve | Write and *solve* one-step first-degree equations, with one variable, involving inverse operations using the set of whole numbers. (SD, gr. 6) |

*Translate* a situation involving multiple arithmetic operations into algebraic form *using* **equations, tables, and graphs.** (WA, gr. 6)

This recurring combination of verbs, concepts, and representations prompted us to organize and analyze the use of verbs within the GLEs. As a result, we identified *99* different verb families (apply, applying, applied, and so on were counted as one verb family) mentioned 5,648 times within the set of algebra-related GLEs. Table 3.6 summarizes the most common

**Table 3.8. An Example of Variation in Levels of Specificity Within GLEs Related to Function**

| *State* | *Grade-level Learning Expectation* |
|---|---|
| Arizona | Describe the rule used in a simple grade-level appropriate function (e.g., T-chart, input/output model). (the same GLE for gr. 4–8) |
| Colorado | In any functional relationship involving whole numbers and common proper fractions, describe how a change in one quantity affects the other. (gr. 5) |
| | In any functional relationship involving positive rational numbers, describe how a change in one quantity affects the other. (gr. 6) |
| | In any functional relationship involving positive rational numbers and integers, describe how a change in one quantity affects the other. (gr. 7) |
| | In any functional relationship involving rational numbers, describe how a change in one quantity affects the other. (gr. 8) |
| Maryland | Complete a function table using a given addition or subtraction rule. (gr. 3) |
| | Complete a function table using a one-operation (+, –, ×, ÷ with no remainders) rule. Assessment limit: Use whole numbers (0–50). (gr. 4) |
| | Complete a one-operation (+, –, ×, ÷ with no remainders) function table. Assessment limit: Use whole numbers or decimals with no more than two decimal places (0–200). (gr. 5) |
| | Interpret and write a rule for a one-operation (+, –, ×, ÷) function table. Assessment limit: Use whole numbers or decimals with no more than two decimal places (0–10,000). (gr. 6) |
| | Describe how a change in one variable in a linear function affects the other variable in a table of values. (gr. 7) |
| | Determine whether functions are linear or nonlinear when represented in words, in a table, symbolically, or in a graph. Assessment limit: Use a graph to determine if a function is linear or nonlinear. (gr. 8) |

verbs used in these five categories of algebra-related GLEs along with the number of appearances of each verb family.

The ways in which these verbs are used vary greatly. Table 3.7 provides examples of GLEs in each substrand highlighting the various uses of the most common verbs.

*Variation among State Documents.* We noted two major areas of variation in the structure of algebra GLEs. The first area of variation is the level of specificity of GLEs. For example, Table 3.8 shows three approaches to specifying GLEs related to function over contiguous grade levels in upper elementary and middle grades. On the one hand, the Arizona document includes the same GLE in each of Grades 4–8 using the words "grade-level appropriate" to differentiate the expectations. The Colorado document includes grade-specific GLEs differenti-

**Table 3.9. Examples of Variation in "Grain Size" of Algebra-Related GLEs**

| *State* | *Grade-Level Learning Expectation* |
|---|---|
| New Hampshire | Demonstrates conceptual understanding of equality by showing equivalence between two expressions (expressions consistent with the parameters of the left- and right-hand sides of the equations being solved at this grade level) using models or different representations of the expressions, solving formulas for a variable requiring one transformation (e.g., $d = rt$; $d/r = t$); by solving multistep linear equations with integer coefficients; by showing that two expressions are not equivalent by applying commutative, associative, or distributive properties, order of operations, or substitution; and by informally solving problems involving systems of equations in a context. (gr. 8) |
| Ohio | Use physical models to add and subtract monomials and polynomials, and to multiply a polynomial by a monomial. (gr. 8) |
| | Use symbolic algebra (equations and inequalities), graphs, and tables to represent situations and solve problems. (gr. 8) |
| | Write, simplify, and evaluate algebraic expressions (including formulas) to generalize situations and solve problems. (gr. 8) |
| | Solve linear equations and inequalities graphically, symbolically, and using technology. (gr. 8) |
| | Solve 2 by 2 systems of linear equations graphically and by simple substitution. (gr. 8) |
| | Interpret the meaning of the solution of a 2 by 2 system of equations (i.e., point, line, no solution). (gr. 8) |
| | Solve simple quadratic equations graphically (e.g., $y = x\hat{}2 - 16$). (gr. 8) |

ated according to the types of numbers used in the functions (e.g., whole numbers, rational numbers). Similarly, the Maryland document includes grade-specific GLEs related to function and also includes specifications regarding assessment limits (i.e., guidance as to what will be included on the state-level assessment for this topic).

In all, 29 state documents specify the types of numbers used in a subset of their GLEs (e.g., *Using whole numbers, complete a function table based on a given rule*. MS, gr. 6) and 28 state documents specified the types of operations in a subset of their GLEs (e.g., *Determine the output for a particular input given a one-operation function rule involving addition, subtraction, or multiplication*. TN, gr. 3). Twenty-one states specify both the types of numbers and types of operations in a subset of their GLEs (e.g., *Write algebraic expressions involving addition or multiplication using whole numbers*. SD, gr. 6).

A second area of variation noted in our analysis involves the number of learning goals specified within a single GLE. For example, some state documents "pack" a single GLE with multiple learning goals, while other state documents include GLEs with a simpler structure, focusing on one goal or idea. Table 3.9 shows examples from two states to illustrate this phenomenon. The GLE from the New Hampshire document is complex in structure and includes multiple learning objectives. The example GLEs from the Ohio document focus on similar ideas but are organized according to multiple GLEs. Given these differences, comparing the number of GLEs across state documents can be misleading.

## ANALYSIS OF ALGEBRA GLES BY SUBSTRANDS

The following sections summarize the methods of analysis and findings related to GLEs within the major Algebra substrands: Patterns; Functions; EEI; Properties; and Relationships between Operations. Each section begins with an overview describing the set of GLEs including their grade-level distribution across K–8. The overview is followed by a summary of the mathematical emphasis of the set of GLEs.

### Patterns

Across the 42 documents, 829 GLEs related to patterns were compiled and considered in the analysis. As noted in Figure 3.2, GLEs related to patterns are often distributed across consecutive grades, chiefly K–5. Thirty-seven of the 42 state documents include at least one GLE related to patterns in Grades K–8. Among those 37 states, 23 include at least one pattern GLE at each grade level, K–8, while six states have pattern GLEs in only Grades 3–5.

For each pattern GLE, we recorded the type of pattern, if specified (e.g., repeating or growing/shrinking patterns). In all, 36 pattern types were identified. We were interested in the number of states that included a particular pattern type at a specific grade level (e.g., Where are growing and shrinking patterns emphasized within the K–8 curriculum across the states?). In order to have a metric that would represent some level of agreement, we took 21 states (half the 42 states) as our benchmark over all grade levels. When we held a higher standard for the core, very few topics made the cut. We included a list of topics that did not make the cut so that the reader can see the full scope of topics in each substrand. Table 3.10 reports pattern topics/concepts that appeared in less than 21 state documents and Table 3.11 reports pattern topics/concepts that appeared in at least 21 state documents along with a breakdown by grade level. Example GLEs for each concept are found in Table 3.12.

**Table 3.10. Pattern Topics/Concepts in Less Than 21 State Documents**

| *Concept* | *Number of States* | *Grade Span* |
|---|---|---|
| Identify missing elements (in a pattern) | 17 | K–8 |
| Variety of patterns | 14 | K–8 |
| Visual/pictorial patterns | 10 | K–8 |
| Using a pattern rule to solve a problem | 5 | K–8 |
| Kinesthetic/motion patterns | 2 | K–8 |
| Created patterns | 6 | K–4 |
| Verbal patterns | 3 | K–3, 5–7 |
| Simple patterns | 14 | K–3, 5,7 |
| Multistep patterns | 4 | K–1, 7-8 |
| Spatial patterns | 2 | K–3 |
| Concrete patterns | 6 | K–3 |
| Ordering objects | 9 | K–2 |
| Identifying objects | 9 | K–2 |
| Attribute blocks | 2 | K–2 |
| Auditory patterns | 1 | K–1 |
| Tessellating patterns | 3 | 1–2 |
| "The Leg Problem"* | 6 | 1–3 |
| Arithmetic sequences | 12 | 2–8 |
| Linear patterns | 14 | 2–8 |
| Iterative patterns | 2 | 2–6, 8 |
| Rotational patterns | 1 | 2 |
| Patchwork patterns | 1 | 2 |
| Non-numeric patterns (explicitly) | 5 | 3–5 |
| Fibonacci sequence | 7 | 4–6, 8 |
| Nonlinear patterns | 7 | 4–8 |
| One-Step patterns | 2 | 6–7 |
| Exponential patterns | 5 | 6–8 |
| Recursive patterns | 5 | 6–8 |
| Triangular patterns | 2 | 7 |
| Handshake problem | 1 | 7 |
| Quadratic patterns | 1 | 8 |
| Algebraic patterns | 1 | 8 |

*Note:* *Reference to problems of the type *number of legs on chair = 4x.*

Some pattern types such as classification and sorting of objects are prominent in the primary grades while other topics such as numeric patterns and pattern rules are emphasized throughout K–8, with particular emphasis in the upper elementary Grades (3–5).

**Table 3.11. Pattern Topics/Concepts in at Least 21 Documents by Grade Level**

| *Concept* | *K* | *1* | *2* | *3* | *4* | *5* | *6* | *7* | *8* |
|---|---|---|---|---|---|---|---|---|---|
| Classification of objects | 20 | 11 | 6 | 1 | – | – | – | – | – |
| Sorting of objects | 26 | 12 | 5 | 2 | – | – | – | – | – |
| Rule/generalization | 6 | 6 | 17 | 20 | 24 | 26 | 22 | 15 | 19 |
| Growing/shrinking patterns | 3 | 7 | 16 | 9 | 6 | 1 | 1 | – | 1 |
| Skip counting | 4 | 13 | 13 | 8 | 2 | 1 | 1 | 1 | – |
| Repeating patterns | 14 | 22 | 19 | 8 | 4 | 1 | 1 | – | – |
| Numeric patterns | 8 | 23 | 17 | 28 | 24 | 22 | 16 | 14 | 12 |
| Geometric figure/shape patterns | – | 3 | 4 | 13 | 9 | 12 | 8 | 3 | 3 |
| Sequences | 6 | 5 | 1 | 5 | 6 | 4 | 4 | 10 | 11 |

**Table 3.12. Pattern Topics/Concepts in at Least 21 Documents With Examples**

| *Concept* | *Examples GLEs* |
|---|---|
| Classification of objects | Sorts and classifies objects by color, shape, size, or kind. (FL, gr. K) |
| Sorting of objects | Sort objects in a set by one attribute such as size, shape, color, or thickness. (MN, gr. K) |
| Rule/ generalization | State the rule that describes a given repeating and growing pattern. (ND, gr. 2) |
| Growing and shrinking patterns | Identify, describe, and extend the rules of multiplicative and growing patterns. (DoDEA, gr. 3) |
| Skip counting | Identify, describe, extend, and create numeric patterns; represent and analyze using skip counting by multiples of 2 and 10 starting with any whole number, and using manipulatives and the 100 chart; represent and analyze numeric patterns using skip counting backward by 10s starting with a multiple of 10, and using manipulatives. (MD, gr. 1) |
| Repeating patterns | Generates repeating patterns for the AB pattern, the ABC pattern, and the AAB pattern; growing (extending) patterns that add 1, 2, and 10 to continue the pattern. (KS, gr. 4) |
| Numeric patterns | Describe and extend simple numeric patterns and change from one representation to another. (MO, gr. 2) |
| Geometric figure/ shape patterns | Knows mathematical relationships in patterns (e.g., the second shape is the first shape turned 90 degrees). (FL, gr. 4) |
| Sequences | Explore and describe patterns and sequences using tables, graphs, and charts. State rules for simple function tables using numbers and ratios. (CT, gr. 3) |

GLEs focused on generalizing patterns and formulating rules based on patterns were included in 39 state documents. The language associated with these GLEs varies by state. For example, students are expected to

**Table 3.13. Uses of a Rule/Generalization for Patterns**

| *Grade* | *Express/Describe a Rule* | *Understand and Apply Rules* | *Analyze Rules* | *Explain/Justify Rules* | *Total* |
|---|---|---|---|---|---|
| K | 6 | 1 | 0 | 0 | 7 |
| 1 | 6 | 0 | 0 | 1 | 7 |
| 2 | 17 | 4 | 0 | 0 | 21 |
| 3 | 20 | 10 | 1 | 0 | 31 |
| 4 | 24 | 8 | 2 | 2 | 36 |
| 5 | 26 | 11 | 0 | 2 | 39 |
| 6 | 20 | 4 | 3 | 0 | 27 |
| 7 | 15 | 3 | 3 | 0 | 21 |
| 8 | 19 | 7 | 1 | 0 | 27 |
| Total | 153 | 48 | 10 | 5 | 216 |

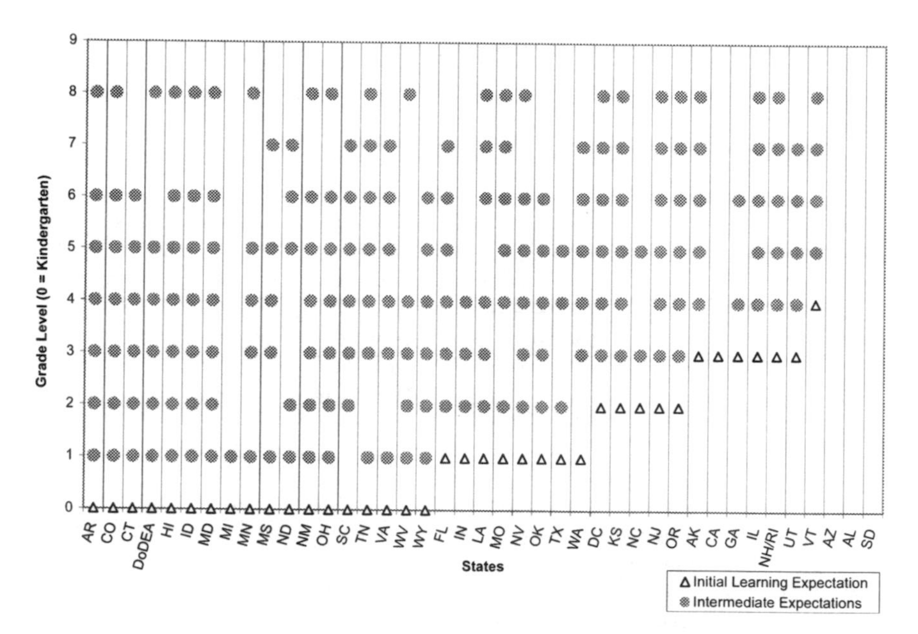


Figure 3.4. Grade placement of GLEs related to pattern rules and generalizations across K–8 by state.

"describe" or "analyze" patterns, "state rules for patterns," or "generalize patterns" throughout Grades K–8. In general, the emphasis in Grades 4–8 appears to be on describing a rule and understanding and applying those rules. The categories expressed in Table 3.13 also suggest levels of cognitive demand. Considering Bloom's Taxonomy as a reference, the

**Table 3.14. Language Used in the Pattern/Rule GLE at the Highest Grade Level of K–8**

| *Term/Phrase* | *Number of States* |
|---|---|
| Analyze patterns | 7 |
| Describe patterns | 8 |
| Generalize patterns | 12 |
| Generalize patterns to the *n*th term | 6 |
| Use of a formula | 2 |
| Formulate symbolic rules | 11 |

expectations move from simply describing the rule through application and analysis to justifying one's reasoning, to validating a rule. When considered in this way, the totals across the bottom of Table 3.13 take on added significance. There are fewer expectations at the higher levels of cognitive demand. Students are asked to *justify* the rule of a pattern in only five GLEs. In contrast, students are asked to *describe* a rule in 153 expectations.

Figure 3.4 summarizes the grade at which pattern rule and generalization GLEs (those noted in Table 3.12) are found in the state documents. The figure notes the earliest (or initial) grade at which such a GLE is found as well as all other grades at which pattern rule/generalization is mentioned.

Table 3.14 provides a summary of the intent of the GLEs at the highest grade level coded in Figure 3.4. This summary provides an indication of common goals related to this topic by the end of elementary/middle school.

## Functions

Only one state made an explicit distinction between relations and functions in their K–8 GLEs, therefore the analysis of the function GLEs included all 611 GLEs from the 42 state documents for relations and functions. There are great differences in the coverage given to functions across the state documents. Twenty-four states have function GLEs in at least five different grade levels while eight states have less than five function GLEs across all of K–8. In states with gaps in their function GLEs, Grade 6 was omitted in seven out of 10 cases.

Concepts related to functions identified in the analysis are summarized in Tables 3.15 and 3.16. Table 3.15 indicates the number of states and the grade span for each of the concepts that appeared in less than 21

**Table 3.15. Function Topics/Concepts in Less Than 21 Documents**

| *Concept* | *Number of States* | *Grade Span* |
|---|---|---|
| $x$-intercept | 7 | 7–8 |
| $y$-intercept | 15 | 7–8 |
| Increasing/decreasing functions | 3 | 6–7 |
| Continuous/discrete data | 2 | 5–8 |
| Functions vs. relations | 1 | 8 |
| Vertical line test | 1 | 8 |
| Domain/range | 1 | 8 |
| Maximum/minimum values | 1 | 8 |
| Direct/inverse proportion | 9 | 4–8 |
| Constant functions | 2 | 6–8 |
| Equation of a line | 10 | 1–8 |
| Quadratic functions | 7 | 7–8 |
| Cubic functions | 4 | 7–8 |
| Exponential functions | 4 | 7–8 |
| Reciprocal functions | 1 | 7–8 |
| Square root functions | 1 | 8 |
| Qualitative change | 9 | K–3 |
| Quantitative change | 9 | K–3, 5, 7 |
| Constant/varying rate of change | 16 | 1–8 |
| Slope as rate of change | 10 | 7–8 |
| Slope as rise over run | 6 | 5, 7–8 |
| Positive/negative slope | 3 | 6–8 |
| Parallel lines/equal slope | 1 | 8 |

**Table 3.16. Function Topics/Concepts in at Least 21 Documents by Grade Level**

| *Concept* | *K* | *1* | *2* | *3* | *4* | *5* | *6* | *7* | *8* |
|---|---|---|---|---|---|---|---|---|---|
| Rule/generalization | – | 4 | 11 | 18 | 21 | 18 | 19 | 20 | 21 |
| Change | 5 | 6 | 12 | 9 | 9 | 15 | 15 | 16 | 14 |
| Independent/dependent variables | – | 1 | 1 | 2 | 6 | 5 | 8 | 19 | 16 |
| Linear functions | – | 1 | 1 | 2 | 3 | 6 | 13 | 29 | 31 |
| Slope | – | – | – | – | – | 1 | 4 | 9 | 23 |
| Nonlinear functions | – | – | – | – | – | 1 | 5 | 10 | 23 |

state documents. Table 3.16 contains grade-level information regarding those concepts that appeared in at least 21 state documents. Example GLEs for each concept are found in Table 3.17.

**Table 3.17. Function Topics/Concepts in at Least 21 Documents With Examples**

| *Concept* | *Example* |
|---|---|
| Rule/generalization | Finds the values and determines the rule with two operations using a function table (input/output machines, T-tables). (KS, gr. 7) |
| Change | Verbally describe changes in various contexts (e.g., plants or animals growing over time). (NM, gr. K) |
| Independent/ dependent variables | Investigate how a change in one variable relates to a change in a second variable. (TN, gr. 4, 5) |
| Linear functions | Select a table of values that satisfies a linear equation and recognize the order pairs on a rectangular coordinate system. (IL, gr. 6) |
| Slope | Use manipulatives and/or appropriate handheld graphing calculators to develop the concept of slope. (AR, gr. 6) |
| Nonlinear functions | Graph simple quadratic equations ($y = kx^2$ or $y = kx^2 + b$) by generating a table of values for a given equation. (OR, gr. 8) |

The concepts in the last 14 rows of Table 3.15 were incorporated into the appropriate categories (Linear Functions, Nonlinear Functions, Change, and Slope) in Table 3.16.

The broad category of generalizations receives attention in Grades 1–8, appearing most significantly in Grades 4 and 8. The category of change receives coverage across all grade levels, however, at no single grade level is it mentioned in half of the states analyzed. Linear functions are emphasized in seventh and eighth grade, while slope and nonlinear functions are primarily eighth-grade topics.

Figure 3.5 summarizes attention to linear functions across the 42 state documents. It indicates little agreement about when functions should first be introduced. That is, several states include GLEs addressing linear functions at every grade level beginning in kindergarten, while other states include linear functions only in the upper elementary/middle grades.

The first mention of a linear function topic occurs as early as kindergarten (five states) and as late as eighth grade (one state), and spans everything between in relatively consistent numbers. Simple relationships between quantities often build the foundation for more advanced study of function. Change is another common topic seen in early grades. For example:

> Verbally describe changes in various contexts (e.g., plants or animals growing over time). (NM, gr. K)

## Expressions, Equations, and Inequalities

Expressions, Equations, and Inequalities (EEI) comprised the greatest number of GLEs (943) within the algebra strand. As a substrand, EEI

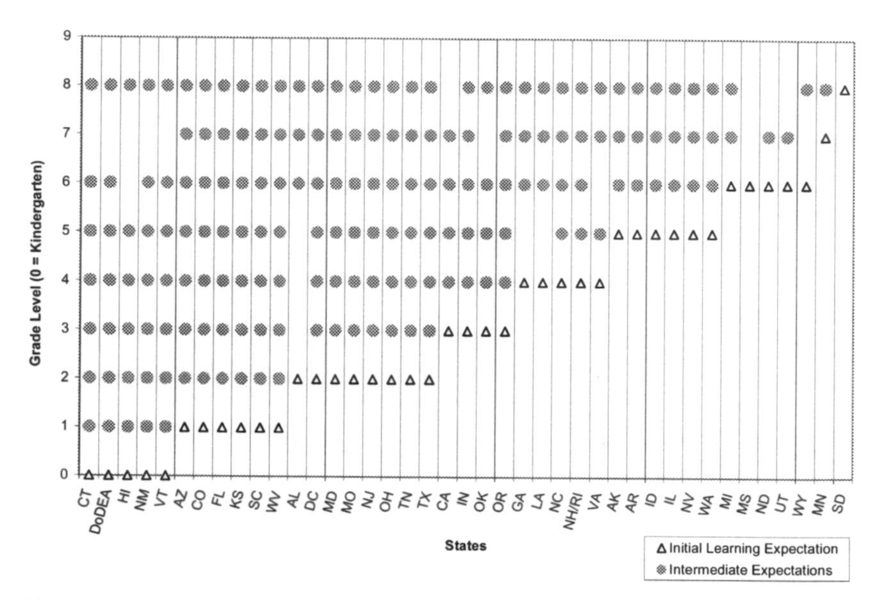


Figure 3.5. Grade placement of GLEs related to linear functions across K–8 by state.

**Table 3.18. Expressions, Equalities, and Inequalities (EEI) Topics/Concepts in Less Than 21 Documents**

| *Concept* | *Number of States* | *Grade Span* |
|---|---|---|
| Proportions | 13 | 3, 6–8 |
| Ratio | 6 | 6–8 |
| Percent | 5 | 6–8 |
| Isolating variables | 2 | 7–8 |
| Multistep equations | 15 | 6–8 |
| Systems of equations | 10 | 8 |
| $D = rt$ | 8 | 4–8 |
| Systems of inequalities | 1 | 8 |
| Expressions/equations with exponents | 12 | 7–8 |
| Operations on monomials/polynomials | 7 | 7–8 |
| Expressions/equations with roots | 6 | 7–8 |
| Quadratic equations | 3 | 8 |
| Expanding $(a+b)^2$, $(a-b)^2$, and $(a+b)(a-b)$ | 1 | 8 |

includes primarily those concepts associated with symbolic algebra. As with Patterns and Functions, the EEI GLEs were coded using a common set of concepts/topics gleaned from multiple readings of the set of GLEs.

**Table 3.19. Expressions, Equations, and Inequalities (EEI) Topics/Concepts in at Least 21 Documents by Grade Level**

| *Concept* | *K* | *1* | *2* | *3* | *4* | *5* | *6* | *7* | *8* |
|---|---|---|---|---|---|---|---|---|---|
| Variables | 1 | 8 | 13 | 15 | 26 | 32 | 23 | 21 | 10 |
| Expressions | – | 1 | 1 | 7 | 13 | 24 | 28 | 39 | 32 |
| Formulae | – | – | – | 2 | 6 | 5 | 13 | 16 | 19 |
| Number sentences/ equations (general) | 2 | 11 | 14 | 21 | 19 | 24 | 26 | 31 | 34 |
| One–Step equations | – | 6 | 11 | 11 | 14 | 12 | 14 | 15 | 9 |
| Two–Step equations | – | – | – | – | – | 5 | 7 | 12 | 16 |
| Inequalities | – | 1 | 2 | 4 | 5 | 9 | 11 | 22 | 28 |

**Table 3.20. EEI Topics/Concepts in at Least 21 Documents With Examples**

| *Concept* | *Example* |
|---|---|
| Variables | Represent the idea of a variable as an unknown quantity, a letter, or a symbol within any whole number operation. (WY, gr. 5) |
| Expressions | Apply the correct order of operations including addition, subtraction, multiplication, division, and grouping symbols to generate equivalent algebraic expressions. (MN, gr. 7) |
| Formulae | Identify information and apply it to a given formula. (SD, gr. 5) |
| Number sentences/ equations | Complete number sentences with missing values and operation symbols. (DC, gr. 2) |
| One-Step equations | Use a variety of methods to model and solve one-step linear equations (e.g., use properties of equality, graph ordered pairs with paper and pencil, use graphing calculators). (OK, gr. 7) |
| Two-Step equations | Solve two-step equations involving whole numbers and a single variable (e.g., $3x + 4 = 19$). (UT, gr. 6) |
| Inequalities | Using whole numbers as a replacement set, find possible solutions to such inequalities as $8 + 4 > n$. (NV, gr. 5) |
| Real-world problems | Justify the selection of a particular value for an unknown quantity in a real-world situation (e.g., Two girls had 10 cookies. If Kwame had 6, how many did Ellie have? Explain). (WA, gr. 2) |

Table 3.18 summarizes those concepts that were mentioned by less than 21 states along with the number of states mentioning the concept and the grade levels at which it appears. Table 3.19 summarizes those concepts that appeared in at least 21 state documents along with the number of states at each grade level. Table 3.20 provides examples of GLEs for each major concept.

Most of the concepts that appeared in less than 21 states were most commonly found in the middle grades. The concepts in the last seven

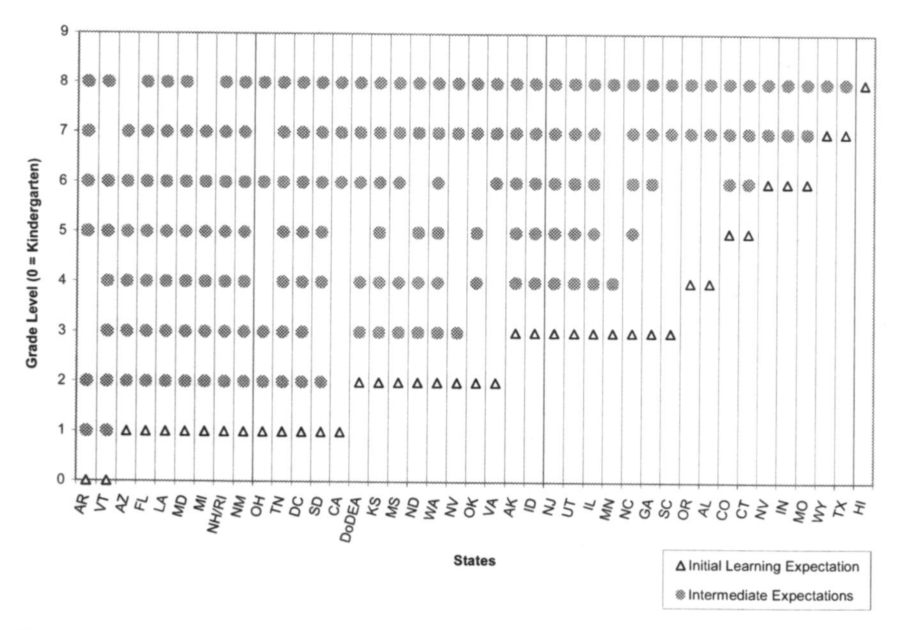


Figure 3.6. Grade placement of GLEs related to linear equation solving across K–8 by state.

rows of Table 3.18 were also included in the appropriate categories in Table 3.19 (Formulae, Inequalities, Expressions, and Equations).

GLEs addressing Variables and Number Sentences/Equations (the general category used when the type of equation was not specified) occur across Grades 1–8. Variables show the most prominence in Grade 5, while Number Sentences and Equations increase across most grades and reach their peak in Grade 8. One-step equations appear in Grades 2–8, however, they do not appear in at least half of the state documents at any single grade level. Expressions occur frequently in four grade levels with a maximum number in Grade 7. Inequalities are present primarily in Grades 7 and 8, increasing in number each year from Grade 1.

Figure 3.6 summarizes attention to linear equations by grade level and state. As noted, the first appearance of GLEs related to linear equation solving ranged between kindergarten and eighth grade. These first experiences, when they occur in early grades, often involve finding the missing addend and/or subtrahend in number sentences. For example:

> Complete addition and subtraction number sentences with a missing addend or subtrahend. (AL, gr. 4)

Some GLEs are very explicit about the type of linear equation solving to be learned. For example:

> Solve two-step equations involving whole numbers and a single variable. (UT, gr. 6)

Other GLEs leave this somewhat open to interpretation. For example:

> Solve simple linear equations and inequalities. (DoDEA, gr. 6)

Some state documents include GLEs related to linear equation solving at every grade level, while other state documents leave gaps at particular grade levels.

## Properties

GLEs related to arithmetic properties (e.g., commutative, associative, and distributive properties) were found in both Number and Algebra strands. In fact, more GLEs involving properties were identified in the Number strand than in the Algebra strand and these were primarily focused on facilitating learning of numerical operations (e.g., commutative property of addition). Overall, there were 158 GLEs related to properties identified from the Algebra strand and 300 from the Number strand. Table 3.21 indicates the properties that were mentioned in less than 21 state documents, while Table 3.22 shows the properties that appeared in at least 21 state documents. Examples for each major concept are provided in Table 3.23.

The data suggest that properties, are most heavily concentrated in the upper elementary to early middle grades. Inverses are primarily emphasized in the middle school Grades (6–8).

## Relationships Between Operations

Most state mathematics standards documents contain GLEs pertaining to relationships between operations and to order of operations. As with

**Table 3.21. Properties in Less Than 21 Documents**

| *Property* | *Number of States* | *Grade Span* |
|---|---|---|
| General properties/unspecified | 12 | 1–8 |
| Additive inverses | 18 | 5–8 |
| Multiplicative inverses | 17 | 5–8 |
| Closure | 2 | 7–8 |

**Table 3.22. Properties in at Least 21 Documents by Grade Level**

| *Property* | *K* | *1* | *2* | *3* | *4* | *5* | *6* | *7* | *8* |
|---|---|---|---|---|---|---|---|---|---|
| Commutative property | 1 | 17 | 17 | 28 | 21 | 17 | 20 | 22 | 17 |
| Identity/zero property | 2 | 10 | 10 | 23 | 15 | 13 | 11 | 15 | 10 |
| Additive identity/zero property | 2 | 11 | 9 | 11 | 5 | 3 | 3 | 6 | 3 |
| Commutative property of addition | 1 | 17 | 13 | 17 | 7 | 5 | 6 | 7 | 3 |
| Associative property | – | 1 | 17 | 18 | 20 | 17 | 15 | 24 | 18 |
| Associative property of addition | – | 1 | 13 | 11 | 6 | 5 | 6 | 4 | 3 |
| Multiplicative identity | – | – | 1 | 19 | 8 | 4 | 5 | 5 | 3 |
| Commutative property of multiplication | – | – | 1 | 24 | 11 | 7 | 6 | 4 | 2 |
| Associative property of multiplication | – | – | – | 12 | 9 | 6 | 6 | 5 | 2 |
| Distributive property | – | – | – | 3 | 14 | 18 | 20 | 21 | 22 |
| Inverse | – | – | – | 2 | 2 | 3 | 8 | 19 | 14 |

Properties, the GLEs for this analysis were extracted only from the Number and Algebra strands. One state did not mention relationships between operations in their K–8 document. Thirty-four of the 42 state documents mention both the inverse relationship between addition and subtraction and the inverse relationship between multiplication and division (often within the same GLE). Twenty-one state documents mention both the relationship between addition and multiplication as well as the relationship between subtraction and division (often within the same GLE). Nineteen states mention all four of these primary relationships. See Tables 3.24 and 3.25 for a summary of the relationships between operations. Table 3.24 includes those relationships mentioned in less than 21 state documents while Table 3.25 summarizes those mentioned in at least 21 states and breaks each topic down by grade level. Examples of GLEs for each major concept are provided in Table 3.26.

GLEs focusing on the relationships between addition and subtraction, multiplication and addition, and division and subtraction are most prominent in Grades 2–4 (2–3, 3, and 3–4, respectively). Although the relationship between multiplication and division is most often seen in Grades 3 and 4, it continues to be mentioned throughout the middle grades. Thirty-nine state documents include GLEs focusing on order of operations most commonly in Grades 4–8, with emphasis in the middle Grades (6–8).

## SUMMARY OF FINDINGS FROM THE ANALYSES OF THE ALGEBRA STRAND

Our main goal in analyzing the Algebra strand in the current state GLE documents was to take a reading on the extent to which tensions over the

**Table 3.23. Properties in at Least 21 Documents with Examples**

| *Property* | *Example* |
|---|---|
| Commutative property | Identify such properties as commutativity, associativity, and distributivity and use them to compute with whole numbers. (A)* (CO, gr. 5) |
| Commutative property of addition | Apply the commutative and identity properties of addition to whole numbers. (N) (MO, gr. 3) |
| Commutative property of multiplication | Identify patterns in related multiplication and division sentences (fact families such as $2 \times 3 = 6$, $3 \times 2 = 6$, $6/2 = 3$, $6/3 = 2$). (A) (TX, gr. 3) |
| Associative property | Demonstrate that changing the order of addends does not change the sum (e.g., $3 + 2 + 7 = 12$, $7 + 3 + 2 = 12$) and changing the grouping of three or more addends does not change the sum (e.g., $[2 + 3] + 7 = 12$, $2 +[3 + 7] = 12$). (A) (UT, gr. 2) |
| Associative property of addition | Demonstrate the associative property of addition (e.g., $[ 3 + 5] + 4 = 3 + [5 + 4]$). (N) (AZ, gr. 2) |
| Associative property of multiplication | Identify specific illustrations of the commutative ($3 + 6 = 6 + 3$) associative ($(3 + 6) + 5 = 3 + (6 + 5)$, and distributive properties ($(5 \times 13) = (5 \times 10) +(5 \times 3)$). (A) (CT, gr. 3) |
| Distributive property | Applies the properties of rational numbers to solve problems (commutative, associative, distributive, identity, equality, inverse). (N) (FL, gr. 7) |
| Additive identity/zero property | Understand the role of zero in addition and subtraction. Example: You start with 6 eggs and then give away 0 eggs. How many eggs do you have now? (N) (IN, gr. 1) |
| Multiplicative identity | Understand and use basic properties of real numbers: additive and multiplicative identities, additive and multiplicative inverses, commutativity, associativity, and the distributive property of multiplication over addition. (A) (MI, gr. 7) |
| Inverse | Use commutative, associative, distributive, identity, and inverse properties to simplify and perform operations. (N) (OH, gr. 5) |

*Note:* * A notation of "A" or "N" indicate whether the GLE is found in the Algebra (A) or Number (N) strand.

development of algebra throughout the curriculum and the placement of a substantive concentration on algebra were in fact being promoted. Here we present the essence of our findings around each of the research questions presented at the beginning of the chapter.

*What are the strands in which algebra expectations are found? What titles do states use to label the algebra grade level expectations for Grades K–8?*

We found algebra expectations articulated mainly in a strand labeled as Algebra, but with varying titles. The words most commonly used in these titles to indicate an Algebra strand were Algebra/Algebraic, Functions,

**Table 3.24. Relationships Between Operations in Less Than 21 Documents**

| *Relationship* | *Number of States* | *Grade Span* |
|---|---|---|
| *Relationships between operations (general) | 8 | 1, 3–5 |
| *Inverse operations (general) | 11 | 3–8 |
| Relationship between squares and square roots | 11 | 7–8 |
| Relationship between cubes and cube roots | 2 | 8 |
| Relationship between the *n*th power and the *n*th root | 1 | 8 |
| Relationship between powers and multiplication | 2 | 6–8 |

*Note:* *The first two categories in the table were only used when the GLE was not specific enough to be included in one of the other categories.

**Table 3.25. Relationships Between Operations in at Least 21 Documents by Grade Level**

| *Topic* | *K* | *1* | *2* | *3* | *4* | *5* | *6* | *7* | *8* |
|---|---|---|---|---|---|---|---|---|---|
| Addition and subtraction as inverse operations | 1 | 5 | 14 | 10 | 6 | 2 | 2 | 2 | 2 |
| Multiplication and division as inverse operations | – | – | 2 | 19 | 18 | 7 | 5 | 9 | 5 |
| Multiplication as repeated addition | – | 1 | 6 | 18 | 5 | 1 | – | 1 | 1 |
| Division as repeated subtraction | – | – | 5 | 10 | 8 | 1 | – | – | – |
| Order of operations | – | – | – | – | 9 | 13 | 26 | 30 | 29 |

**Table 3.26. Relationships Between Operations in at Least 21 Documents With Examples**

| *Relationships Between Operations* | *Example* |
|---|---|
| Addition and subtraction as inverse operations | Show relationships between addition and subtraction using physical models, diagrams, and acting-out problems. (WA, gr. 1, 2) |
| Multiplication and division as inverse operations | Use the inverse relationship between multiplication and division to compute and check results. (CA, gr. 3) |
| Multiplication as repeated addition | Represent multiplication as repeated addition. (IL, gr. 3, 5) |
| Division as repeated subtraction | Describe models of sharing equally (division) as repeated subtractions and arrays. (SC, gr. 2) |
| Order of operations | Explore the order of operations. (ID, gr. 6) |

Patterns, Relations/Relationships, and Reasoning/Thinking (Table 3.1). Other expectations that fit our Algebra coding-scheme were found in other strands, most notably Number.

*What labels typically are used in state GLEs to guide the development of the algebra substrands, K–8?*

The most common words used in labels to indicate algebra substrands were Patterns, Algebra/Algebraic, Function/Functional, Relations/Relationships, Models/Modeling, Equations/Equality, Symbol/Symbolic, Change, Representations, Expressions, and Inequalities (Table 3.2). The *PSSM* (NCTM, 2000) substrand titles were also used by seven states. We identified five substrands for our analysis, Patterns, Functions, EEI (Expressions, Equations, and Inequalities), Properties, and Relationships between Operations.

*Within each identified substrand what are major topics/concepts on which the states have agreement at particular grade levels?*

There were no concepts or topics in algebra for which all 42 states at a given grade level include an expectation specific enough to code for the concept or topic. The greatest agreement reflected in our analysis is that 39 of the 42 states include an expectation that students should study algebraic expressions in Grade 7. The next highest level of agreement is that 32 states expect students to study variables at Grade 5 and expressions at Grade 8 (Table 3.19). The major result from our analysis is the *lack of agreement* on what should be expected at each grade level in the substrands of algebra.

*What are typical grade levels or grade bands over which a substrand has major emphasis?*

For Patterns, the major emphasis is in Grades K–4 with less emphasis in upper elementary and middle school. Both EEI (Expressions, Equations, and Inequalities) and Functions have their major emphasis in the middle Grades 6–8. Properties and Operations have a peak of emphasis in Grade 3 followed by a decline over Grades 4 and 5. The greatest emphasis is reached for both Properties and Relationships between Operations at Grade 7 (Figures 3.1–3.3).

*What is the scope of differences in grade placements for learning objectives related to algebra? What is typical?*

In algebra few topics reach mastery over Grades K–8. However, there is ample evidence that states vary substantially in the grade levels at which they concentrate on particular algebra topics. For example, the levels at which states expect the commutative property of multiplication to be taught vary from Grade 2 to 8, with Grades 3 and 4 having the greatest concentration of states (Table 3.22). The levels at which states expect knowledge of variables ranges from kindergarten to Grade 8 (Table 3.19). In general, the EEI and Function GLEs are concentrated at Grades 6 to 8, while Patterns are concentrated at the lower grades.

While these state documents have expectations of algebra concepts in lower grade levels, the migration is not as apparent as the rhetoric in the United States would imply. There is a gradual buildup to more symbolic algebra at Grades 7 and 8, but the work at the lower grades seems to be more conceptual with gradual exposure to ideas.

*What is the range of levels of specificity of GLEs?*

The specificity expressed in standards across states ranged considerably. Wording such as "simple" or "appropriate" and "grade-level appropriate" were sometimes used to specify the same general content over several consecutive grade levels. Interpretation is left to the teacher or test constructor to determine what is appropriate. However, there were other states that were very specific in their goals for students at each of the grade levels.

Table 3.8 gives an example contrasting the Arizona, Colorado, and Maryland approach to specifying GLEs. In the Function category, Arizona includes the same GLE in each of Grades 4–8, using the words "grade-level appropriate" to differentiate the expectations. Colorado includes a greater level of specificity by differentiating according to the types of numbers used in the functions (e.g., whole numbers, rational numbers). Maryland goes a step further by specifying not only the types of numbers being used, but also the operations included at each grade level. We found such differences were common across the various states.

Another difference between expectations across states is the sheer amount of content expressed in an expectation (Table 3.9). One state may have a single expectation at a particular grade level, but that expectation is the equivalent of six expectations at the same grade level in another state. The number of "objectives" in a particular expectation could express varying numbers of actions to be taken as well as different kinds of expectations. In other cases, related ideas that were stand-alone expectations in some states were combined into a single expectation in others.

*What constitutes the agreed-upon algebra expectations for K–8 for the greatest cluster of states?*

Our analysis indicated little overall agreement across documents in the algebra expectations for a particular grade level. Therefore, in order to

**Table 3.27. Algebra Topics/Concepts in at Least 21 of 42 State Documents Analyzed**

| | | |
|---|---|---|
| Patterns | Classification of objects<br>Sorting of objects<br>Rule/generalization<br>Growing and shrinking patterns<br>Patterns involving skip counting | Repeating patterns<br>Numeric patterns<br>Geometric figure/shape patterns<br>sequences |
| Functions | Rule/generalization<br>Change<br>Independent/dependent variables | Linear functions<br>slope<br>Nonlinear functions |
| Expressions, Equations, and Inequalities (EEI) | Variables<br>Expressions<br>Formulae<br>Number sentences/equations | One–step equations<br>Two–step equations<br>Inequalities |
| Properties | Commutative property of addition<br>Commutative property of multiplication<br>Associative property of addition<br>Associative property of multiplication | Distributive property<br>Additive identity<br>Multiplicative identity<br>Inverse (additive and multiplicative) |
| Relationships between Operations | Addition and subtraction as inverse operations<br>Multiplication and division as inverse operations | Multiplication as repeated addition<br>Division as repeated subtraction<br>Order of operations |

have a metric that would represent some level of agreement, we took 21 states (half the 42 states) as our benchmark over all grade levels. When we held a higher standard for the core, very few topics made the cut. With 21 states as the cut point for inclusion, we found the core of topics/concepts of algebra contained in the K–8 state mathematics standards as specified in Table 3.27.

Table 3.27 does not tell the whole story, but it does give a picture of the algebra concepts on which at least 21 states agree should be taught somewhere in Grades K–8. As we summarized above, the Pattern work is emphasized in the early grades with Function and EEI (Expressions, Equations, and Inequalities) being more prominent in the middle Grades (6–8). Work with Properties and Relationships between Operations is most concentrated in Grade 7. Although this table presents some agreement among states on major ideas, the grade placement of these topics varies greatly.

CHAPTER 4

# ANALYSIS OF EMPHASIS ON REASONING IN STATE MATHEMATICS CURRICULUM STANDARDS

**Ok-Kyeong Kim and Lisa Kasmer**

Reasoning mathematically is fundamental to learning mathematics with understanding. When reasoning is effectively promoted and fostered in the classroom through predicting and justifying results and making sense of observed phenomena, students develop a deeper understanding and connection of mathematical ideas. In turn, these understandings and connections help students reason mathematically, which can result in improved learning outcomes. Ball and Bass (2003) state that mathematical learning cannot be considered without reasoning. The National Council of Teachers of Mathematics (NCTM; 2000) argues,

> Reasoning and proof should be a consistent part of students' mathematical experience in pre-kindergarten through Grade 12. Reasoning mathematically is a habit of mind, and like all habits, it must be developed through consistent use in many contexts. (p. 56)

*The Intended Mathematics Curriculum as Represented in State-Level Curriculum Standards: Consensus or Confusion?* 89–109

Reasoning is a broad and general term and is used in different ways by researchers, curriculum developers, and teachers. For example, *PSSM* (NCTM, 2000), which includes reasoning and proof as one of the five process standards, emphasizes the importance of making and investigating mathematical conjectures; developing and evaluating mathematical arguments; and selecting and using various types of reasoning and methods of proof. The *TIMSS* provided an assessment framework on reasoning that included the following elements: (1) hypothesize/conjecture/predict, (2) analyze, (3) evaluate, (4) generalize, (5) connect, (6) synthesize/integrate, (7) solve nonroutine problems, and (8) justify/prove (Mullis, Martin, Smith, et al., 2001). Stylianides and Silver (2004) developed a framework to analyze mathematics curricula in terms of reasoning. Their framework focused on the process of proving, that is, identifying a pattern, making a conjecture, providing a proof, and providing a nonproof argument. Ball and Bass (2003) view reasoning as a process of inquiry and a process of justification. The former is used for "discovering and exploring new ideas" and the latter is used for "justifying and proving mathematical claims" (p. 30).

Our goal in this analysis is to describe the emphasis within state GLE documents related to reasoning. The framework for the analysis is based on a review of the literature as well as a preliminary analysis of the state mathematics curriculum documents. In our initial review, three broad categories of reasoning surfaced: *reasoning as conceptual understanding, reasoning used in problem solving,* and *reasoning for verification*. First, reasoning can be considered as a form of conceptual understanding. As such, reasoning is about making meaning, concept development, connections among concepts, and relationship building. Second, reasoning is used in various phases of problem solving including: (1) analyzing problem situations, (2) developing and applying strategies, (3) selecting and applying strategies and mathematical ideas, (4) explaining strategies, and (5) checking the reasonableness of the results in the problem context. Finally, reasoning can be considered as a thought process through which students make and test conjectures and prove or disprove them. This includes justification, verification, and validation.

For our analysis of the state standards documents, we selected *reasoning for verification* as our primary focus. This allowed us to narrow the scope of the work. Our goal was to summarize the emphasis on reasoning in GLEs from kindergarten to eighth grade. In doing so, we describe the extent and nature of emphasis on reasoning in five content areas (i.e., Number and Operations, Algebra, Geometry, Measurement, and Data Analysis and Probability), how GLEs related to reasoning are organized in the documents, and overall characteristics of reasoning expectations across documents.

## METHODOLOGY

In this study we analyzed 35 state standards documents, that is, all of the state-level GLE documents available as of December, 2004, that were official (not draft form). (See Appendix A for a complete list of documents reviewed for this analysis.) A cursory analysis of the state documents involved extracting overall characteristics of each state document, such as whether the document included a separate reasoning strand as well as reasoning aspects in content areas, whether it stated explicitly the importance of reasoning, and whether reasoning was integrated throughout the document. This analysis enabled us to document overall features across state documents as well as some specific features that were observed in a few of them.

The major emphasis of our analysis focused on describing the set of GLEs pertaining to reasoning for verification. Initially, GLEs from 15 states were reviewed to guide and develop a draft framework. Based on our examination of the 15 state documents and frameworks from the literature review, we determined differentiating aspects of GLEs related to reasoning for verification. Keywords determined to be our core modifiers included predict, generalize, verify, validate, justify, support, defend, refute, conclusions, inferences, conjectures, hypotheses, arguments, claims, test, prove, disprove, examples/counterexamples, evaluate, and inductive/deductive. However, these key words were not the sole means of identifying the set of GLEs to be analyzed. Ultimately, decisions were made based on the context and content of individual GLEs in each document. Once the preliminary framework was determined, the remaining state standards were reviewed and the researchers refined the framework.

Each state mathematics standards document was reviewed independently by two researchers to identify all GLEs that related to one or more of the categories of reasoning identified for this analysis. Subsequent to the independent reviews, the researchers met to develop a final data set, reconciling discrepancies. Following the initial evaluation of the GLEs, those for reasoning were consolidated and cross-referenced.

Using the initial set of keywords, reasoning GLEs from the 35 state documents were compiled. Then, they were reviewed once more to determine whether each GLE supported the aspects of reasoning (i.e., reasoning for verification) and cross-checked with the original documents reviewed in order to make sure that the GLEs that were compiled matched the documents. Once a set of reasoning GLEs was compiled by keyword, similar aspects were grouped together; for example, making conjectures and making hypotheses. The groups were: (1) prediction, (2) generalization, (3) verify/validate, (4) justify, (5) draw conclusions or make inferences, (6) make conjectures or hypotheses, (7) test conjectures, hypotheses, or predictions, (8) develop arguments, (9) prove or disprove/refute, (10) evaluate claims, hypotheses, predictions, conclusions, or conjectures, (11) use examples/counterexamples, and (12) refine or modify conjectures.

For each group of GLEs, a plot using a form consisting of each state on the horizontal axis and grade level on the vertical axis was constructed (see figures in the results section). The plots as well as the organized data set were searched for common patterns and themes.

## FINDINGS

This section provides a summary of the organization of GLEs related to reasoning for verification. It also includes a summary of the attention to this topic by content strand, grade level, and state.

### Organization of Learning Expectations

Fourteen state GLE documents (40%) have a strand on reasoning in addition to GLEs within content strands that focus on reasoning, although not all of the GLEs within the reasoning strand are grade specific. For instance, four states (Alaska, Arizona, Idaho, and Washington) contain a strand on reasoning at each grade level and five states (Minnesota, Nevada, Oklahoma, Texas, and Utah) have a reasoning section unique for each grade band. Five states (Georgia, New Jersey, Oregon, South Carolina, and Department of Defense Education Activity) present the same reasoning section or expectation in all grades. In addition, 15 states include a general statement that reasoning is expected throughout content strands. To summarize, 22 states (63%) have either a separate reasoning section or an explicit statement that reasoning should be embedded in all content strands.

Our analysis noted some negative features related to attention to reasoning for verification, including inconsistency of message within documents (across grade levels), isolation of reasoning from content, inappropriate use of examples, and lack of clarity of GLEs.

*Inconsistency.* Even though states appear to recognize the importance of reasoning, there is inconsistency in emphasis on reasoning across grades and across content areas. There are very few reasoning GLEs in the Number and Operation strand or in the Measurement strand. However, many detailed reasoning GLEs are found in the Data Analysis and Probability strand. In addition, there are a limited number of reasoning GLEs in primary grades. For example, 25 states (71%) have limited, if any, GLEs related to reasoning in primary grades.

There is also a discrepancy between components of state curriculum documents when addressing reasoning for verification. For example, Hawaii has sections delineated as "benchmarks" and "performance indicators" to address GLEs. Some of the benchmark statements do not specify reasoning aspects, but their corresponding performance indicators support reasoning. An example is seen in Table 4.1.

In other states, some reasoning GLEs are not supported by their lower-level GLEs. For example, New Mexico geometry in Grades 5–8 has an expectation (named as "benchmark"): *Analyze characteristics and properties of two- and three-dimensional geometric shapes and develop mathematical arguments about geometric relationships,* but the specific GLEs (named as "performance standards") in each of grades from 5–8 do not support the development of ability to make mathematical arguments (see Table 4.2).

*Isolation from content.* State standards documents with a separate reasoning section/strand tend not to specify or link reasoning GLEs to content strands. Instead, reasoning GLEs tend to be broad and general, and isolated from specific content. For example:

**Table 4.1. An Example From Hawaii's Benchmarks and Performance Indicators**

| *Benchmarks 6–8* | *Grade 6 Performance Indicator* |
|---|---|
| • Select appropriate units to estimate and measure angles, circumferences and areas of a circle, and surface area and volume of regular solids.<br>• Estimate and measure angles in plane figures.<br>• Develop and use formulas for circumference and area of a circle.<br>• Develop and use formulas to find surface area and volumes of regular solids.<br>• Select appropriately precise tools to achieve a desired accuracy in measurement.<br>• Use ratios and proportions to solve problems related to measurement.<br>• Determine an appropriate scale and make scale drawings; use scale drawings or models in applications.<br>• Solve simple problems involving rates and other derived measures (e.g., miles per hour, gms per cc or $cm^3$). | The student:<br>1. Determines and ***justifies*** appropriate units needed to estimate and measure angles.<br>2. Gives an estimate between 10° above or below the actual measurement of a given angle.<br>3. Uses a protractor to determine actual angle measurements to the nearest degree.<br>4. Selects and uses appropriate standard measuring tools (e.g., graduated cylinder, measuring scales, trundle wheel, measuring tape) to achieve desired accuracy in measurement.<br>5. Determines and ***justifies*** the accuracy of measurement needed for a problem. |

**Table 4.2. New Mexico Geometry GLEs (Grades 5–8)**

| *5–8 Benchmark: Analyze characteristics and properties of two- and three-dimensional geometric shapes and* ***develop mathematics arguments*** *about geometric relationships* | |
|---|---|
| *Grade* | *Performance Standards* |
| 5 | 1. Identify, describe, and classify two-dimensional shapes and three-dimensional figures by their properties.<br>2. Recognize and describe properties of regular polygons having up to 10 sides.<br>3. Identify faces, edges, and bases on three-dimensional objects. |
| 6 | 1. Identify, describe, and classify the properties of, and the relationships between, plane and solid geometric figures:<br>•Measure, identify, and draw angles, perpendicular and parallel lines, rectangles, and triangles by using appropriate tools (e.g., straightedge, ruler, compass, protractor, drawing software)<br>•Understand that the sum of angles of any triangle is 180 degrees and the sum of the angles of any quadrilateral is 360 degrees and use this information to solve problems<br>•Visualize and draw two-dimensional views of three-dimensional objects made from rectangular solids<br>2. Classify angles as right, obtuse, or straight.<br>3. Describe the properties of geometric figures that include regular polygons, circles, ellipses, cylinders, cones, spheres, and cubes.<br>4. Classify polygons as regular or irregular.<br>5. Classify triangles as scalene, isosceles, or equilateral and by angles (i.e., right, acute, and obtuse).<br>6. Identify angle, line, segment, and ray and use the symbols for each.<br>7. Describe the relationship between radius, diameter, and circumference of a circle. |
| 7 | 1. Classify geometric figures as similar or congruent.<br>2. Understand the concept of a constant (e.g., pi) and use the formulas for the circumference and area of a circle.<br>3. Explain and use the Pythagorean theorem.<br>4. Determine the radius, diameter, and circumference of a circle and explain their relationship.<br>5. Use properties to classify solids including pyramids, cones, prisms, and cylinders. |
| 8 | 1. Recognize, classify, and discuss properties of all geometric figures including point, line, and plane.<br>2. Identify arc, chord, and semicircle and explain their attributes.<br>3. Use the Pythagorean theorem and its converse to find the missing side of a right triangle and the lengths of the other line segments. |

> Formulate conjectures and discuss why they must be or seem to be true. (ID, gr. 6)

In particular, the Minnesota state document indicates that the mathematical reasoning standard (one of five strands around which all of the GLEs are organized) will primarily be assessed within the context of the standards in the remaining four strands (Number Sense, Computation, and Operations; Patterns, Functions, and Algebra; Data Analysis, Statistics, and Probability; and Spatial Sense, Geometry, and Measurement) and yet this document provides very few reasoning GLEs in content strands (see figures in the Results section).

*Inappropriate use of examples.* As indicated earlier, some state documents provide examples to illustrate the intent of GLEs. Some of these examples do not support reasoning, however. Moreover, some reasoning GLEs contain examples that focus only on procedures. For example, Kansas has the following GLE and example in seventh-grade number and operations:

> Explains the mathematical reasoning that was used to solve a real-world problem using a one- or two-step linear equation. For example, Kim has read 5 more than twice the number of pages as Hank. Kim has read 15 pages. How many pages has Hank read? To solve, write $2h + 5 = 15$. Then to find the answer subtract 5 from both sides of the question. $2h = 10$, then divide both sides of the equation by 2, so $h = 5$. (KS, gr. 7)

The GLE requires reasoning and yet the example includes only procedural explanation. On the other hand, some state documents include examples that address reasoning even though the stated GLE does not specify an emphasis on reasoning. For example,

> Represent two-variable data with a scatter plot on the coordinate plane and describe how the data points are distributed. If the pattern appears to be linear, draw a line that appears to best fit the data and write the equation of that line. Example: Survey some of the students at each grade level in your school, asking them how much time they spend on homework. Plot the grade level and time of each student as a point (grade, time) on a scatter diagram. Describe and justify any relationship between grade and time spent on homework. (IN, gr. 8)

The GLE does not address reasoning directly, but the example requires students to demonstrate reasoning.

*Variations in clarity.* Various levels of specificity and clarity are evident in the reasoning for verification GLEs of the state standards documents. Some GLEs are very specific and simple in their structure; others are simple in structure but lack clarity. Therefore, simplicity of a GLE does not necessarily imply clarity. On the other hand, some GLEs are conveyed with lots of words (and examples) and yet they may or may not convey a clear message. Table 4.3 presents examples of GLEs in various categories related to specificity and clarity.

**Table 4.3. Example GLEs Related to Reasoning**

| *Category* | *Example* |
|---|---|
| Simple structure, vague meaning | • Analyze and interpret data (prediction, inference, conclusion, etc.). (AR, gr. 4, Data Analysis and Probability)<br>• Make simple predictions based on a variable (e.g., a child's height from year to year). (AZ, gr. 2, Algebra)<br>• Develop and evaluate inferences and predictions that are based on models. (WV, gr. 3, Data Analysis and Probability) |
| Simple structure, specific/clear meaning | • Skip-count on a hundreds chart to identify, describe, and predict number patterns. (NM, gr. 1, Algebra)<br>• Verify the Pythagorean theorem using an area dissection argument. (AZ, gr. 8, Geometry)<br>• Predict which of two events is more likely to occur if an experiment is repeated. (VA, gr. 2, Data Analysis and Probability) |
| Complex structure, vague meaning | • Explain the relationship of numbers in one- and two-dimensional graphs (e.g., number lines and coordinate graphs), with and without appropriate technology such as graphing calculators. (AR, gr. 5, Number and Operation) |
| Complex structure, specific/clear meaning | • Estimate from a graph or a set of data the mean and standard deviation of a normal distribution and draw conclusions about the distribution of data using measures of center and spread. (OR, gr. 8, Data Analysis and Probability)<br>• Analyze and make predictions about patterns involving whole numbers, decimals, and fractions using a variety of tools including organized lists, tables, objects, and variables. (UT, gr. 5, Algebra)<br>• In response to a teacher- or student-generated question or hypothesis, collects appropriate data, organizes the data, appropriately displays/represents numerical and/or categorical data, analyzes the data to draw conclusions about the questions or hypothesis being tested, and when appropriate makes predictions, asks new questions, or makes connections to real-world situations. (VT, gr. 5, Data Analysis and Probability) |

## Grade-Level Learning Expectations on Reasoning Within Content Strands

In this section, we describe the emphasis on reasoning within the content strands in the 35 state curriculum documents. As noted in Chapter 1, process strands (e.g., problem solving, reasoning, communication) are not common across state documents and specification of learning expectations related to reasoning by grade level are even less common. Therefore, our focus is on GLEs related to reasoning within content strands (i.e., Number and Operations, Algebra, Geometry, Measurement, and Data Analysis and Probability).

Words and phrases commonly used to convey learning expectations related to reasoning include predict, generalize, justify/verify, make and test conjectures/hypotheses, draw conclusions/inferences, develop and evaluate arguments. The most common is "predict" or "prediction." Other words associated with reasoning are used much less frequently. Examples include prove, disprove, validate, support/defend, defend/refute, provide evidence, inductive/deductive, and revise conjectures. In the following sections we summarize the emphasis on reasoning by examining GLEs that use the following set of words/phrases commonly associated with the topic:

- Predict;
- Generalize;
- Verify;
- Justify;
- Draw Conclusions/Make Inferences;
- Make Conjectures;
- Test Conjectures;
- Develop Arguments; and
- Evaluate Claims.

*Prediction.* GLEs related to prediction are found in most of the content strands of state standards documents, although Data Analysis and Probability contain the greatest number of predication GLEs. Examples from each content strand are provided below along with states, grade levels, and content strands.

> Predict what comes next in an established pattern and justify thinking. (UT, gr. K, Algebra)
>
> Use a sample space to predict the probability of an event. (TN, gr. 5, Data Analysis and Probability)
>
> Predict and describe the results of sliding, flipping, and turning two-dimensional shapes. (NM, gr. 3, Geometry)
>
> Make and test predictions about measurements, using different units to measure the same length or volume. (DoDEA, gr. 2, Measurement)
>
> Use estimation to predict or to verify the reasonableness of calculated results. (WA, gr. 8, Number and Operation)

Figure 4.1 provides a summary of the grades at which states include GLEs related to prediction by state. Note that every state has at least one GLE pertaining to prediction. A few states (e.g., Florida, Idaho, and New Mexico) include prediction GLEs at every grade level. Eight states (Department of Defense Education Activity, District of Columbia, Florida, Idaho, New Mexico, Texas, Utah, and Virginia) provide prediction GLEs in the kindergarten level and 17 states in Grade 1, whereas more than 20 states have 'prediction' GLEs in each grade from Grades 2–8.

Even though state documents include expectations of making predictions, they generally do not provide follow-up expectations, such as testing or evaluating predictions and comparing predictions to results. For example, only 10 out of the 35 states (29%) have an expectation of testing predictions (see Figure 4.7). Figure 4.9 provides evidence that evaluating predictions is not included in most state documents. Fourteen states (Arizona, District of Columbia, Department of Defense Education Activity, Hawaii, Missouri, New Mexico, Ohio, Oklahoma, Oregon, South Carolina, Tennessee, Utah, Virginia, and West Virginia) have one or two expectations such as verifying, justifying, or supporting predictions. Six states (Arizona, Department of Defense Education Activity, New Mexico, Oregon, Virginia, and Washington) have an expectation of comparing predictions to results. Most of these expectations are in the strand of data analysis and probability and they do not appear consistently across grades.

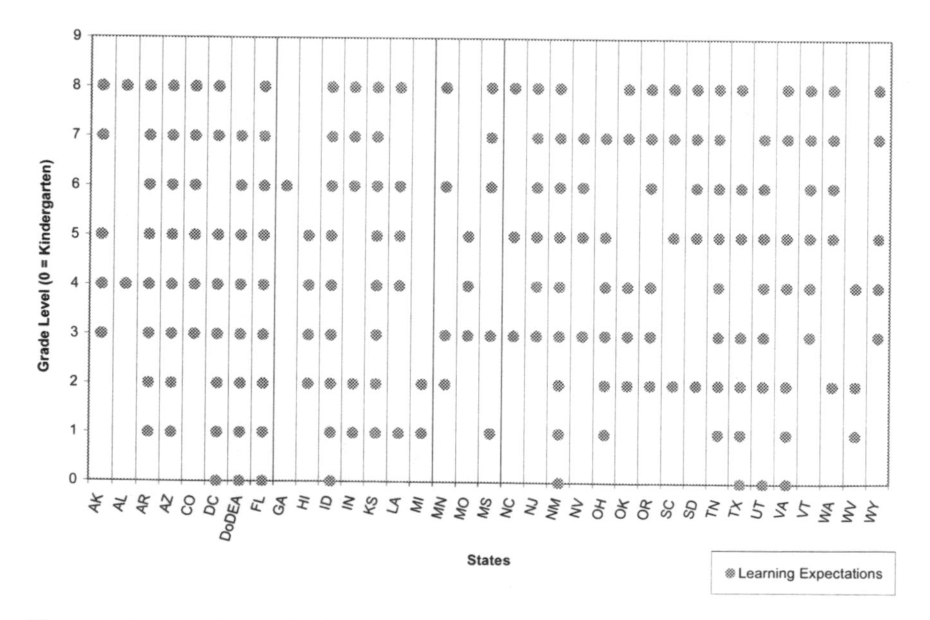


Figure 4.1. Grades at which at least one GLE related to *Prediction* is found, by state.

*Generalization.* GLEs related to generalization are concentrated within the Algebra strand of most (21) state documents. Very few are found in other content strands. Examples of generalization GLEs include:

> Formulate and record generalizations about patterns in a variety of situations (e.g., addition and subtraction patterns, showing the cost of one pencil at 10 cents and two pencils at 20 cents). (OK, gr. 2, Algebra)
>
> Form generalizations and validate conclusions about properties of geometric shapes including parallel lines, perpendicular lines, bisectors, triangles, and quadrilaterals. (NV, gr. 8, Geometry)
>
> Analyze and develop generalizations of exponential patterns, including zero as an exponent, using manipulatives and calculators. (OK, gr. 7, Number and operations)
>
> Recognize, describe, extend, create, and generalize patterns by using manipulatives, numbers, and graphic representations. (WY, gr. 4, Algebra)

Figure 4.2 provides a summary of the grades at which states include GLEs related to generalization by state. Twenty-one of the 35 states (60%) have at least one GLE regarding generalization. As noted, many state

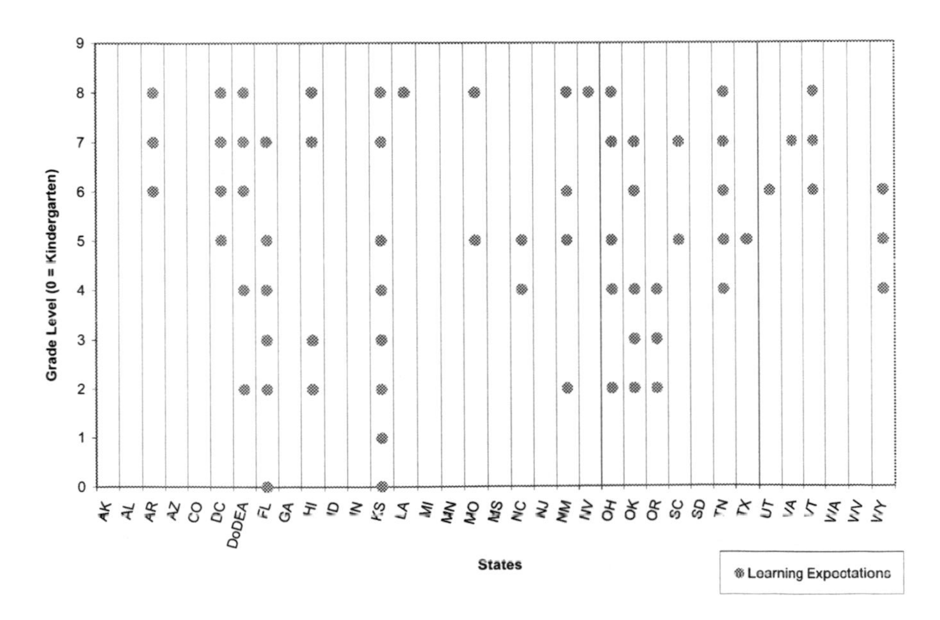


Figure 4.2. Grades at which at least one GLE related to *Generalization* is found, by state.

documents include no reference to generalization within the Grades K–8 GLEs. On the other hand, a few states (Department of Defense Education Activity, Florida, Kansas, New Mexico, Ohio, Oklahoma, and Tennessee) include some emphasis on generalization across multiple grades.

*Verification*. GLEs related to verification are found in most of the content strands of state standards documents, although they are concentrated in two strands: Geometry and Number and Operation. Examples from each content strand are provided below along with states, grade levels, and content strands.

> Verify symmetry by drawing lines of symmetry in shapes and objects. (DoDEA, gr. 3, Geometry)
>
> Estimates and verifies by counting sets that have more, fewer, or the same number of objects (for example, using a reference set of objects, comparing cards with different numbers of dots, estimating whether sets are more or less than a given number such as five). (FL, gr. K, Number and Operation)
>
> Write and solve linear equations and inequalities in one variable, interpret the solution or solutions in their context, and verify the reasonableness of the results. Example: As a salesperson, you are paid $50 per week plus $3 per sale. This week you want your pay to be at least $100. Write an inequality for the number of sales you need to make, solve it, and check that your answer is reasonable. (IN, gr. 8, Algebra)

Figure 4.3 provides a summary of the grades at which states include GLEs related to verification by state. Twenty-one state documents (60%) have GLEs pertaining to verification. Most of the 21 states include GLEs related to verification in the upper grades and a very limited number in primary grades. Students are asked to verify at various levels or degrees of reasoning. For example, several verification expectations ask students to use particular tools or methods to verify results, such as calculators, measuring tools, counting, drawing lines of symmetry, and using tracing procedures. Other types of verification include verifying predictions, conclusions, results, solutions, mathematical relationships, or mathematical ideas.

*Justification*. For this analysis, we identified all GLEs that included any form of the words justify, support, or defend. GLEs related to justification are found in each of the content strands of state standards documents, with the greatest attention in Data Analysis and Probability. Figure 4.4 provides a summary of the grades at which states include GLEs related to

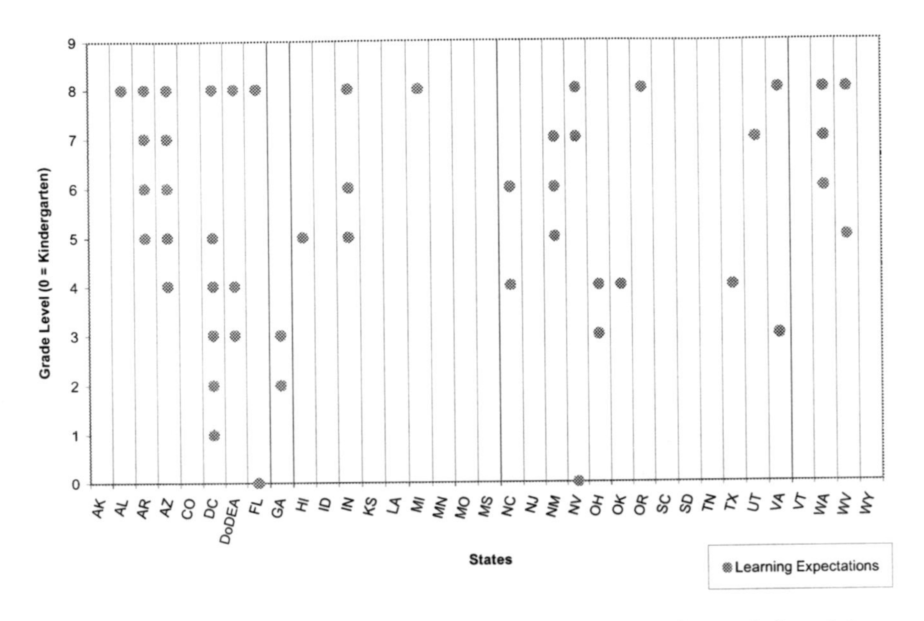


Figure 4.3. Grades at which at least one GLE related to *Verification* is found, by state.

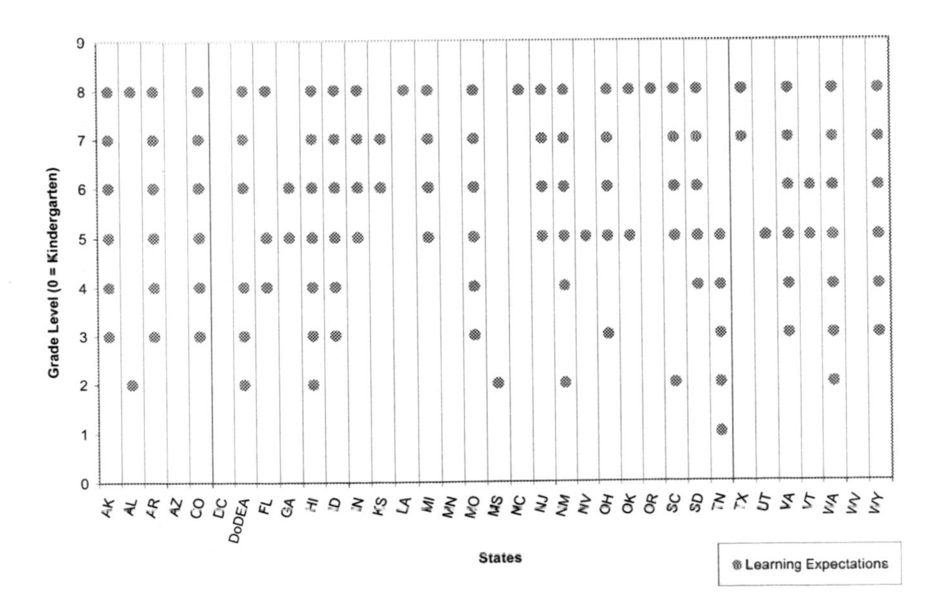


Figure 4.4. Grades at which at least one GLE related to *Justification* is found, by state.

justification by state. Examples of justification GLEs are provided below along with states, grade levels, and content strands.

> Justifying the strategy used to solve addition and subtraction problems. (AL, gr. 2, Number and Operation)
>
> Applies statistical data to predict trends and to make and justify generalizations. (FL, gr. 5, Data Analysis and Probability)
>
> Uses logical arguments to justify conclusions about properties of lines in the plane. (HI, gr. 4, Geometry)

All but four states (Arizona, District of Columbia, Minnesota, and West Virginia) have at least one GLE regarding justification. In most state documents, attention to justification begins after the primary grades. In fact, five states (Kansas, Louisiana, North Carolina, Oregon, and Texas) include their first GLE focused on justification at Grades 6, 7, or 8. An analysis of the GLEs reveals that students are expected to justify at various levels or degrees of reasoning. For example, in some GLEs students are asked to justify choices. These include justifying the choice of a particular representation or display (Data Analysis and Probability), selection of tool (Measurement), and choice of unit of measure (Measurement). Other GLEs expect students to justify reasonableness of an estimate or measurement, or reasonableness of a result. These expectations are evidenced primarily in the strand of Number and Operation. Other justification GLEs focus on justifying solutions, conclusions, arguments, reasoning, generalizations, strategies, predictions, inferences, or ideas.

*Draw conclusions/make inferences.* GLEs related to drawing conclusions and making inferences are concentrated in the Data Analysis and Probability strand. Examples include:

> Read, interpret, and draw conclusions from various displays of data. (AK, gr. 5, Data Analysis and Probability)
>
> Draw conclusions and answer questions using information organized in real-object graphs, picture graphs, and bar-type graphs. (VT, gr. 1, Data Analysis and Probability)
>
> Using models, make and test conjectures about geometric properties and relationships and explain the conclusions. (SC, gr. 3, Geometry)
>
> Model problem situations and draw conclusions using representations such as graphs, tables, or number sentence. (MO, gr. 5, Algebra)
>
> Uses the mathematical modeling process to analyze and make inferences about real-world situations. (KS, gr. 8, Algebra)

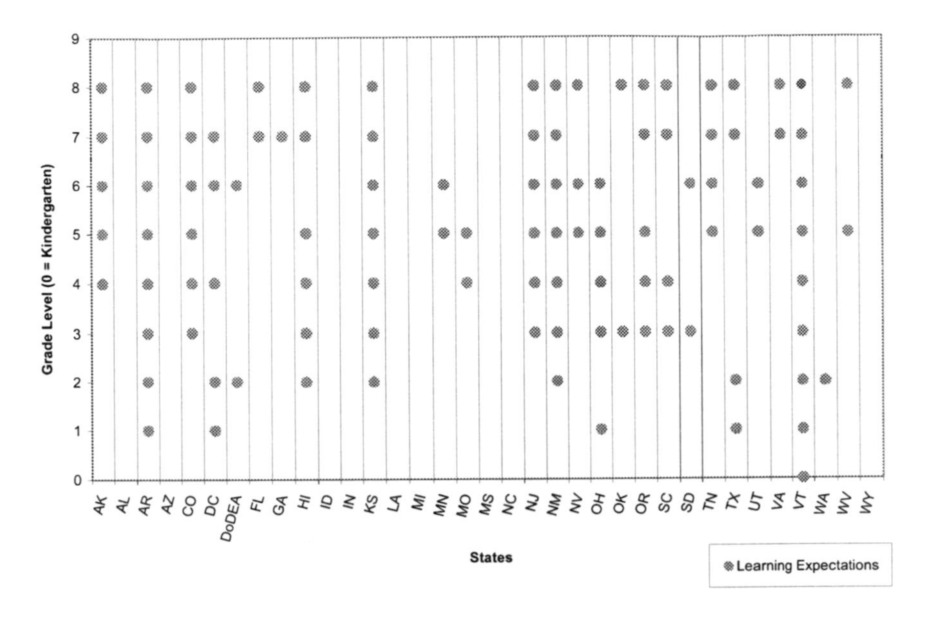


Figure 4.5. Grades at which at least one GLE related to *Making Inferences and/or Conclusions* is found, by state.

Figure 4.5 provides a summary of the grades at which states include GLEs related to making inferences and/or conclusions by state. As shown, 26 states (74%) have GLEs referencing drawing conclusions or making inferences. Six states (Arkansas, Colorado, Kansas, New Jersey, New Mexico, and Vermont) include GLEs focused on drawing conclusions or inferences across most, if not all, the grades, and these tend to be within the strand of data analysis and probability.

*Make conjectures*. GLEs related to making conjectures are concentrated in the Geometry and Data Analysis and Probability strands. Examples include:

> Use observations about differences between two or more samples to make conjectures about the populations from which the samples were taken. (OR, gr. 7, Data Analysis and Probability)
>
> Use logic and reasoning to make and support conjectures about geometric objects. (NJ, gr. 7, 8, Geometry)
>
> Develop, test, and explain conjectures about properties of whole numbers and commonly used fractions and decimals.
> (CO, gr. 5, Number and Operation)

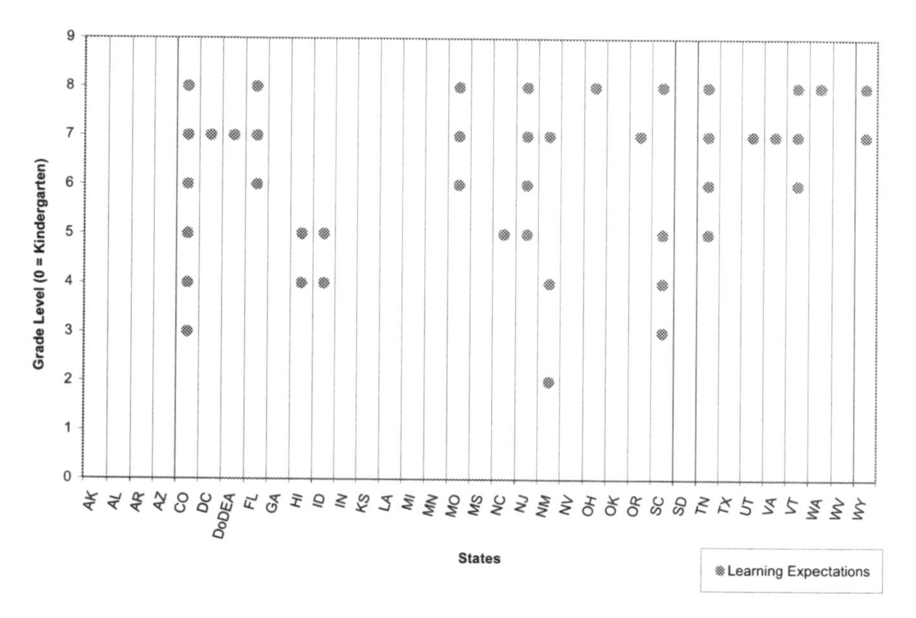


Figure 4.6. Grades at which at least one GLE related to *Making Conjectures* is found, by state.

> Generates new questions or hypotheses as a result of data collected. (HI, gr. 4, Data Analysis and Probability)

Figure 4.6 provides a summary of the grades at which states include GLEs related to making conjectures by state. Nineteen of the 35 states (54%) include GLEs regarding making conjectures or hypotheses. In eight of the 19 states (Department of Defense Education Activity, District of Columbia, North Carolina, Ohio, Oregon, Utah, Virginia, and Washington), making a conjecture or hypothesis appears only in one grade. Most conjecture expectations appear in Grades 5–8 and seldom in primary grades.

*Test conjectures.* GLEs related to testing conjectures are concentrated in the Geometry and Data Analysis and Probability strands. Examples include:

> Test the validity of properties by using examples of the properties of operations on real numbers. (VA, gr. 7, Number and Operations)
>
> Makes and tests conjectures about geometric properties and relationships and forms conclusions based on tests. (HI, gr. 5, Geometry)

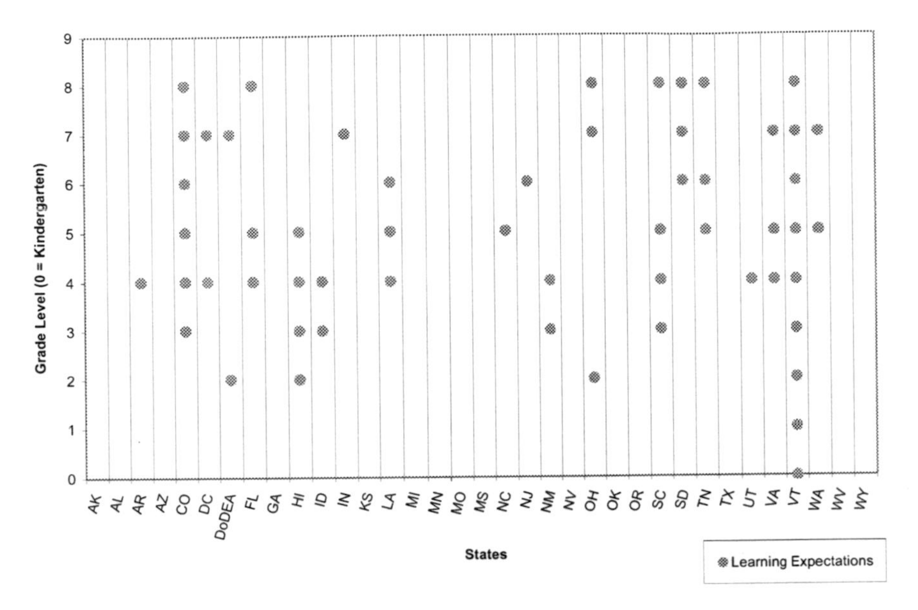


Figure 4.7. Grades at which at least one GLE related to *Testing Conjectures* is found, by state.

> Make and test predictions about measurements, using different units to measure the same length or volume. (OH, gr. 2, Measurement)
>
> Develop and test conjectures about properties of integers (Does 3 – 5 = 5 – 3?) and rational numbers. (CO, gr. 8, Number and Operation)
>
> Formulate a hypothesis and then design and carry out an experiment to test it. (TN, gr. 8, Data Analysis and Probability)

Twenty state documents (57%) include GLEs about testing conjectures, hypotheses, or predictions (see Figure 4.7). As with making conjectures, Data Analysis and Probability is the content strand where most of these GLEs occur. Half of the 20 state documents include GLEs focused on testing conjectures in only one or two grades and most are found in Grades 4–8. The Colorado and Vermont documents include the greatest attention to testing conjectures.

*Develop arguments.* Fourteen states (40%) include GLEs related to developing arguments (see Figure 4.8), and these primarily appear in the upper grades and within the Data Analysis and Probability strand. Examples include:

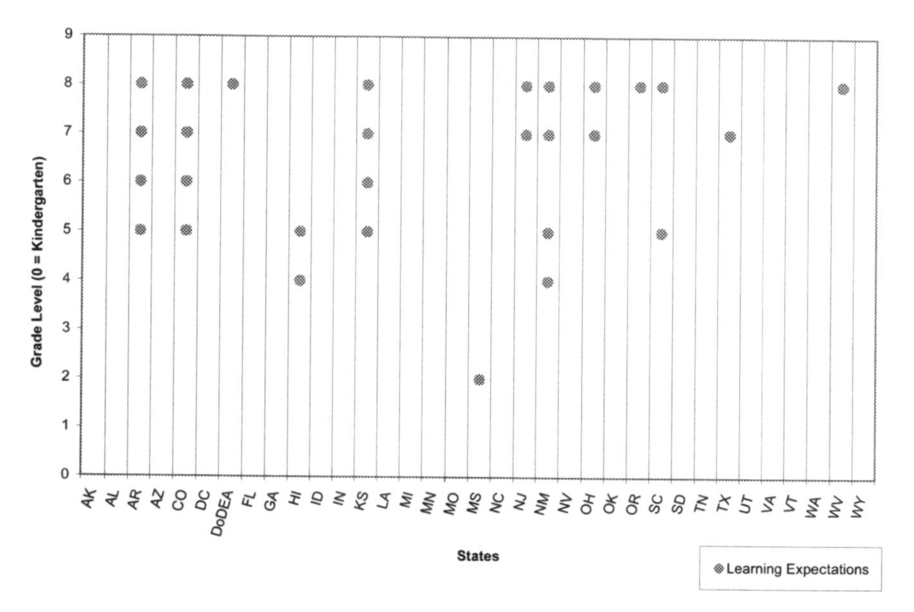


Figure 4.8. Grade at which at least one GLE related to *Developing Arguments* is found, by state.

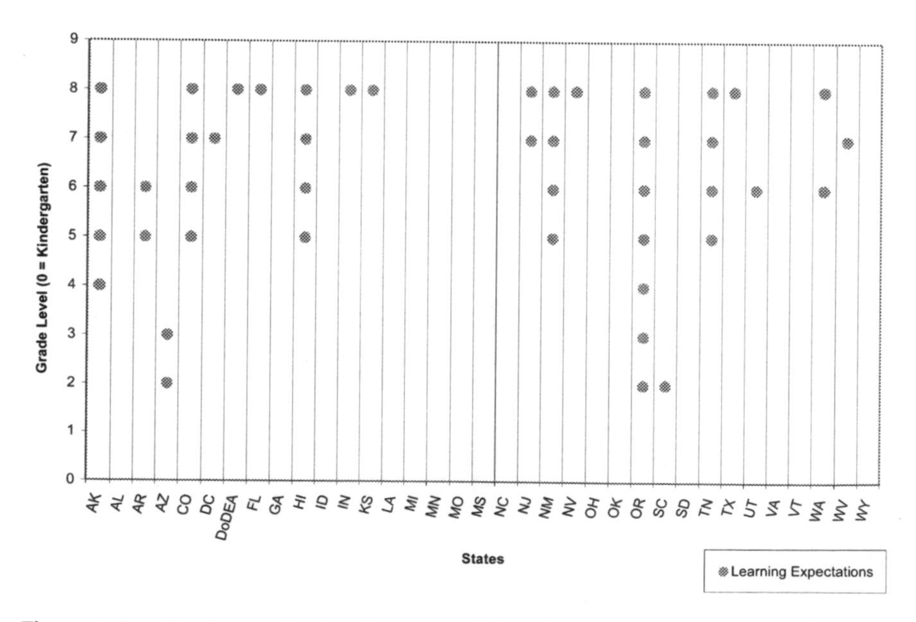


Figure 4.9. Grades at which at least one GLE related to *Evaluating Arguments* is found, by state.

> Make and test conjectures about geometric relationships and develop logical arguments to justify conclusions. (CO, gr. 5, Geometry)
>
> Use convincing arguments to justify the selection of a specific unit of measure for a given item. (MS, gr. 2, Measurement)
>
> Identify claims based on statistical data and in sample cases, evaluate the validity and usefulness of the claims. (NM, gr. 6, Data Analysis and Probability)
>
> Make inferences and convincing arguments based on an analysis of given or collected data. (TX, gr. 7, Data Analysis and Probability)

*Evaluate claims.* Twenty states include GLEs focused on evaluating claims, hypotheses, arguments, conjectures, conclusions, inferences, predictions, or strategies. Once again, they are concentrated in the Data Analysis and Probability strand at the upper grade levels (see Figure 4.9). Examples include, all of which are found in the strand of Data Analysis and Probability:

> Evaluate arguments that are based on statistical data. (AR, gr. 5–6)
>
> Recognize faulty arguments or common errors in data analysis. (DoDEA and KS, gr. 8)
>
> Evaluates the hypothesis by making inferences and drawing conclusions based on statistical results. (FL, gr. 8)
>
> Predict and evaluate how adding data to a set of data affects measures of center. (OR, gr. 7)
>
> Develop and evaluate inferences and predictions based on data. (UT, gr. 6)

Comparing Figures 4.5 and 4.9, it is noteworthy that states with expectations of making conclusions or inferences do not have expectations of evaluating conclusions or inferences. There are only two states that provide consistent expectations in both: Alaska and Tennessee. Comparing Figure 4.8 with Figure 4.9, it is found that the fact that a state has an expectation of developing arguments does not guarantee that this state also has an expectation of evaluating arguments. Only Colorado and New Jersey require both expectations. Figures 4.6, 4.7, and 4.9 illustrate that an expectation of making conjectures or hypotheses is not offered with an expectation of testing or evaluating them. Only Colorado and South Carolina have consistent expectations of making and testing conjectures or hypotheses to some degree.

**Table 4.4. Number of State Curriculum Documents That Include GLEs in Each Category of the Reasoning Framework**

| *Reasoning Focus* | *K* | *1* | *2* | *3* | *4* | *5* | *6* | *7* | *8* |
|---|---|---|---|---|---|---|---|---|---|
| Prediction | 8 | 17 | 24 | 24 | 24 | 26 | 22 | 25 | 27 |
| Generalization | 2 | 1 | 8 | 5 | 9 | 10 | 10 | 12 | 12 |
| Verification | 2 | 1 | 2 | 5 | 7 | 7 | 6 | 6 | 13 |
| Justification | 1 | 1 | 8 | 12 | 14 | 23 | 20 | 19 | 24 |
| Conclusion/inference | 1 | 6 | 9 | 12 | 13 | 16 | 15 | 16 | 17 |
| Conjecture | 0 | 0 | 1 | 2 | 5 | 7 | 6 | 13 | 10 |
| Testing | 1 | 1 | 4 | 6 | 12 | 10 | 6 | 9 | 7 |
| Argument | 0 | 0 | 1 | 0 | 2 | 6 | 3 | 7 | 11 |
| Evaluation | 0 | 0 | 3 | 2 | 2 | 7 | 9 | 9 | 14 |

**Table 4.5. Number of State Documents That Include GLEs in Each Content Strand in the Categories of the Reasoning Framework**

| | *Number* | *Algebra* | *Geometry* | *Measurement* | *Data/Probability* |
|---|---|---|---|---|---|
| Prediction | 10 | 19 | 21 | 4 | 32 |
| Generalization | 1 | 21 | 3 | 0 | 2 |
| Verification | 12 | 7 | 12 | 2 | 5 |
| Justification | 15 | 14 | 13 | 8 | 20 |
| Conclusion/inference | 0 | 7 | 3 | 0 | 26 |
| Conjecture | 2 | 0 | 10 | 0 | 17 |
| Testing | 4 | 0 | 10 | 2 | 15 |
| Argument | 0 | 0 | 6 | 1 | 10 |
| Evaluation | 2 | 1 | 2 | 2 | 17 |

## SUMMARY

It is evident that state curriculum documents from kindergarten to Grade 8 recognize the importance of reasoning. They include various learning expectations related to reasoning, in some cases providing supporting examples to clarify the intent of the GLE. However, four of the set of GLEs related to reasoning for verification suggests inconsistency across the state documents.

Tables 4.4 and 4.5 summarize the number of state curriculum documents that address reasoning aspects by grade and by content strand, respectively. Table 4.4 indicates that reasoning for verification receives minimal attention in the primary grades and that the strand of Data Analysis and Probability has the largest concentration of GLEs related to reasoning. These tables, along with Figures 4.1–4.9, suggest that reasoning expectations are provided in state curriculum documents with great variation in terms of grade, content strand, and state.

The inconsistencies noted across state documents suggest there is no agreed-upon core curriculum with regard to reasoning. Prediction expectations are the most prevalent among the reasoning expectations, followed by justification expectations while making arguments, proving, disproving, and using counterexamples to refute claims have less attention in state curriculum documents. It is not clear that state curriculum documents present a coherent, graduated treatment of reasoning over any particular span of Grades, K–8.

Overall, state curriculum documents acknowledge the significance of reasoning in the mathematics curriculum and try to incorporate reasoning in their GLEs in various ways. Many, however, fail to address reasoning aspects in a thorough and comprehensive manner. It is not evident that reasoning GLEs in the state curriculum documents are consistent across grades and content strands. In addition, reasoning GLEs are not presented with other related reasoning GLEs. For example, many states have "make arguments" without "evaluate arguments" or vice versa. Some state documents also have components that do not match each other when addressing reasoning expectations. Finally, the levels of clarity and specificity of reasoning GLEs vary across states and even within states.

CHAPTER 5

# RECOMMENDATIONS FOR FUTURE DEVELOPMENT OF K–8 MATHEMATICS CURRICULUM STANDARDS

**Barbara J. Reys and Glenda Lappan**

Current educational reform efforts such as the federal No Child Left Behind initiative rely on accountability measures including annual student assessments in mathematics to both prod school improvement initiatives and to document progress in meeting student learning goals. Schools must demonstrate student progress in meeting learning goals in order to avoid federally-imposed sanctions. Thus, because they are the basis of the annual grade-level assessments, mathematics learning goals as described in official state-level mathematics curriculum standards documents have become a central feature of school improvement efforts. According to respondents of a recent survey of state-level curriculum leaders, most teachers and school administrators are paying close attention to the curriculum standards provided by state education agencies, more so than to previous state-level curriculum standards. In fact, over two thirds of respondents perceived that the newest state-level curriculum

*The Intended Mathematics Curriculum as Represented in State-Level Curriculum Standards: Consensus or Confusion?* 111–115

standards are significantly influencing classroom instruction, textbook selection, and professional development for teachers (Reys, Dingman, Sutter, Teuscher, 2005).

Our review of state-level mathematics curriculum standards was prompted, in part, by a desire to understand the range of learning goals at particular grade levels across the state documents. Findings from this study confirm that mathematics learning expectations vary along several dimensions including grain size (e.g., level of specificity), language used to convey learning goals (e.g., understand, explore, memorize and so on), and the grade placement of specific learning expectations. In particular, while the set of learning expectations related to specific topics are similar across the K–8 state standards, the grade placement for any particular topic or learning expectation varies considerably. That is, state-level standards documents generally include similar goals for learning in the K–8 number and algebra strands. However, when the topics are introduced, their trajectory of development across grades and the grade at which students are expected to be fluent with particular mathematical content differs dramatically across the states.

Considerable effort at the state level has led to increased specificity regarding mathematics learning goals across the K–8 years of schooling. However, while national documents such as those produced by NCTM and Achieve, Inc. have served as a common reference document (Reys et al., 2005), few states are collaborating on local work and they are interpreting the national recommendations differently. A notable exception is a collaborative effort by a consortium of New England states (Vermont, Maine, New Hampshire, and Rhode Island) in the development of grade level expectations and/or related student assessments.

As described in this report, most state-level mathematics standards documents are unique in design, organization and grade-placement of learning expectations. As states look ahead to the next cycle of review and revision of mathematics standards, it is likely that they will continue to look to national groups for curriculum leadership. In fact, 85 percent of respondents indicated that national leadership is needed to assist in future articulation of curriculum standards in mathematics, particularly from national professional organizations of mathematics teachers (K–12 and university) and mathematicians (Reys et al., 2005). The two reasons most often cited for increased national leadership were: to increase the level of expertise and resources in developing a well-articulated mathematics curriculum and to promote higher, yet appropriate, learning expectations.

In the next section we offer recommendations for the future development and/or revision of mathematics curriculum standards. The recommendations are based on our review of the current state-level documents and include attention to the organization of the documents as well as to the content and specification of learning expectations within the documents. We also propose more collaboration among states in the development and refinement of GLEs for mathematics so that greater consensus can be achieved.

## RECOMMENDATIONS

The development and implementation of curriculum standards in mathematics is a periodic process. At the state level in the United States and at the federal level in other countries, curriculum standards are reviewed and revised about every 6–10 years. These revisions are prompted by a desire to stay current with available research on how students learn as well as adapt to changing societal and workplace expectations and mathematical developments. As states plan for the next cycle of review, we offer here some suggestions.

- *Identify major goals or focal points at each grade level, K–8.* At each grade, we recommend a general statement of major goals for the grade. These general goals may specify emphasis on a few strands of mathematics or a few topics within strands. These general goals should be coordinated across all grades, K–8, to ensure curricular coherence and comprehensiveness. Offering these major goals will provide guidance to teachers in appropriation of instructional time. It may also help reduce superficial treatment of many mathematical topics, a common criticism of the United States mathematics curriculum. For guidance in determining the major goals or focal points, we suggest collaboration across states and/or guidance from groups such as NCTM and Achieve, Inc.
- *Limit the number of grade-level learning expectations to focus instruction and deepen learning.* The set of learning expectations per grade level should be manageable given the school year. Along with the statement of general goals and priorities for a particular grade, we suggest that the set of learning expectations per grade be limited to 20–25. This number is similar to curriculum standards documents in other countries and may help authors of standards documents develop an appropriate grain size for communicating learning expectations.
- *Organize K–8 grade-level learning expectations by strand.* We recommend that mathematics curriculum standards be organized by grade and by content strand. Further, we recommend that attention be given to both content strands (e.g., Number and Operation,

Geometry, Measurement, Algebra, Data Analysis and Probability) and important mathematical processes (e.g. Problem Solving, Reasoning, and Representations).

- *Develop clear statements of learning expectations focusing on mathematics content to be learned*. We recommend that learning expectations be expressed succinctly, coherently, and with optimum brevity, limiting the use of educational terms (jargon) that may not communicate clearly to the intended audience of teachers, school leaders, and parents. Learning expectations should focus on the mathematics to be learned rather than pedagogy to be employed in presenting the mathematics. The set of learning expectations for a grade should include mathematics to be learned at that grade level (not just what will be assessed).
- *Limit the use of examples within learning expectations*. Some states GLE documents include examples (occasionally or frequently, depending on the document) to clarify the learning expectations and others do not. In some documents, the examples also include messages regarding suggested pedagogy. We recommend limited use of examples within statements of grade-level learning expectations. Instead, we urge authors to strive for clarity within the statement of the learning expectation. If additional information and/or guidance is needed for specific audiences (e.g. teachers or parents), we suggest that a supplement (or companion document) be developed for this particular purpose.
- *Involve people with a broad spectrum of expertise*. Many different constituent groups have valuable knowledge and expertise to contribute to the development of mathematics curriculum standards. These groups include: classroom teachers, mathematicians, curriculum supervisors, and researchers in the fields of mathematics education and cognitive psychology. We recommend that all voices be heard and taken into account in the development of appropriate and rigorous mathematics curriculum standards.
- *Collaborate to promote consensus.* Fifty states with 50 state standards documents increases the likelihood of large textbooks that treat many topics superficially. In order to increase the likelihood of focused curriculum materials, states will need to work together to create some level of consensus about important learning goals and expectations at each grade. This can be accomplished through state consortiums such as the New England Consortium mentioned earlier, through collaborative efforts sponsored by groups such as the National Council of Supervisors of Mathematics, the Association of State Supervisors of Mathematics or the Council of Chief State

School Officers. It can also be accomplished if states can build their curriculum standards from a "core curriculum" offered by national groups such as the NCTM, the College Board and/or Achieve, Inc. In fact, we recommend that a consortium of national groups collaborate to propose a national core curriculum that focuses on priority goals for each grade, K–8. In this way, states might still tailor their own curriculum goals and expectations around local needs while ensuring a much greater level of consistency across the states.

Clearly much work and effort has occurred at the state level related to articulating learning expectations for Grades K–8 mathematics. The state-level GLE documents present specific learning goals and also describe learning sequences for attaining these goals across the elementary years of schooling. For many states, grade-level learning expectations represent a new level of state leadership for curriculum articulation. Although individual documents may provide increased clarity and coherence within their respective state, as a collection they highlight a consistent lack of national consensus regarding common learning expectations in mathematics at particular grade levels.

We have offered some recommendations for future development of mathematics curriculum standards documents to help improve the coherence, continuity, and articulation of mathematics curriculum offered to United States students. It will take strong leadership, cooperation, and collaboration to realize the goal of a coherent, rigorous mathematics curriculum for all United States students. There is no better time to begin this work.

# REFERENCES

Ball, D. L., & Bass, H. (2003). Making mathematics reasonable in school. In J. Kilpatrick, W. G. Martin, & D. Schifter (Eds.), *A research companion to Principles and Standards for School Mathematics* (pp. 27–44). Reston, VA: National Council of Teachers of Mathematics.

Blank, R., & Pechman, E. (1995). *State curriculum frameworks in mathematics and science: How are they changing across the states?* Washington, DC: Council of Chief State School Officers.

Blank, R. K., Langesen, D., Bush, M., Sardina, S., Pechman, E., & Goldstein, D. (1997). *Mathematics and science content standards and curriculum frameworks: States progress on development and implementation*. Washington, DC: CCSSO. Retrieved July 31, 2005, from http://www.ccsso.org/publications/details.cfm?PublicationID=66

Cambridge Conference on School Mathematics. (1963). *Goals for school mathematics*. Boston: Houghton Mifflin for Educational Services Incorporated.

Conference Board of the Mathematical Sciences National Advisory Committee on Mathematical Education. (1975). *Overview and analysis of school mathematics Grade K–12*. Washington, DC: National Council of Teachers of Mathematics.

Flanders, J. R. (1987). How much of the content in mathematics textbooks is new? *Arithmetic Teacher, 35*(1), 18–23.

Ginsburg, A., Leinwand, S., Anstrom, T., & Pollock, E. (2005). *What the United States can learn from Singapore's world-class mathematics system (and what Singapore can learn from the United States): An exploratory study*. American Institutes for Research. Retrieved July 31, 2005, from http://www.air.org/news/documents/Singapore.htm

*The Intended Mathematics Curriculum as Represented in State-Level Curriculum Standards: Consensus or Confusion?* 117–122

Grouws, D., & Smith, M. S. (2000). Finding from NAEP on the preparation and practices of mathematics teachers. In P. Kenney & E. Silver (Eds.), *Results from the seventh mathematics assessment of the national assessment of educational progress* (pp. 107–140). Reston, VA: NCTM.

Klein, D., Braams, B. J., Parker, T., Quirk, W., Schmid, W., & Wilson, W. S. (2005). *The state of state Math standards*. Washington, DC: Thomas B. Fordham Foundation.

Long, V. M. (2003). The role of state government in the custody battle over mathematics education. In G. Stanic & J. Kilpatrick (Eds.). *A history of mathematics education* (pp. 931–954). Reston, VA: National Council of Teachers of Mathematics.

Martin, W. G., Stein, M. K., & Ferrini-Mundy, J. (2002). The impact of *Principles and Standards for School Mathematics* on state policies and practices: Executive Summary. Reston, VA: National Council of Teachers of Mathematics.

McKnight, C. C., Crosswhite, F. J., Dossey, J. A., Kifer, E., Swafford, J. O., Travers, K. J. & Cooney, T. J. (1987). *The underachieving curriculum.* Champaign, IL: Stipes.

Mullis, I., Martin, M., Gonzalez, E., O'Connor, K., Chrostowski, S., Gregory, K., Garden, R., & Smith, T. (2001). *Mathematics benchmarking report TIMSS 1999 – Eighth grade: Achievement for U.S. states and districts in an international context.* Chestnut Hill, MA: TIMSS International Study Center, Boston College.

Mullis, I., Martin, M., Smith, T., Garden, R., Gregory, K., Gonzalez, E., Chrostowski, S., & O'Connor, K. (2001). *TIMSS assessment frameworks and specifications 2003.* Chestnut Hill, MA: TIMSS International Study Center, Boston College.

National Committee on Mathematical Requirements. (1923). *Reorganization of mathematics in secondary education*. New York: Mathematical Association of America.

National Commission on Excellence in Education. (1983). *A nation at risk: The imperative for educational reform*. Washington, DC: U.S. Government Printing Office.

National Council of Teachers of Mathematics. (1980). *An agenda for action*. Reston, VA: Author.

National Council of Teachers of Mathematics. (1989). *Curriculum and evaluation standards for school mathematics*. Reston, VA: Author.

National Council of Teachers of Mathematics. (2000). *Principles and standards for school mathematics*. Reston, VA: Author.

National Education Association. (1899). *Report of the committee on college-entrance requirements.* In Journal of proceedings and addresses of the thirty-eighth annual meeting held at Los Angeles California, July 11-14, 1899 (pp. 632-677). Chicago: Author.

National Institute of Education. (1976). *Conference on basic mathematical skills and learning (Euclid, Ohio): Contributed Position Papers. Vol. 1 and Working Group Reports. Vol. 2.* Washington, DC: Author.

National Research Council. (2002). Curriculum as a channel of influence: What shapes what is taught to whom? In I. R. Weiss, M. S. Knapp, K. S. Hollweg, & G. Burrill (Eds.), *Investigating the influence of standards: A framework for research in mathematics, science, and technology education* (pp. 39–47). Washington, DC: National Academy Press.

No Child Left Behind Act. (2001). Public law no. 107-110. Retrieved January 13, 2005, from http://www.ed.gov/policy/elsec/leg/esea02/index.html

Porter, A. C. (2004). *Curriculum assessment*. Unpublished manuscript. Retrieved August 10, 2006, from http://www.wcer.wisc.edu/scalemsp/research/Products/Porter_CurriculumAssessment.pdf

Porter, A. C., & Smithson, J. L. (2001). *Defining, developing, and using curriculum indicators*. CPRE Research Report (RR-048). Retrieved August 21, 2005, from http://www.cpre.org/Publications/rr48.pdf

Reys, B. J., Dingman, S., Sutter, A., & Teuscher, D. (2005). *Development of state-level mathematics curriculum documents: Report of a survey*. Retrieved July 31, 2005, from http://mathcurriculumcenter.org/news.html

Schmidt, W. H. (2004). A vision for mathematics. *Educational Leadership, 61*(5), 6–11.

Schmidt, W. H., McKnight, C. C., Valverde, G. A., Houang, R. T., & Wiley, D. (1997). *Many visions, many aims* (Vol. 1). Boston: Kluwer.

Stylianides, G., & Silver, E. (2004). Reasoning and proving in school mathematics curricular: An analytic framework for investigating the opportunities offered to students. In D. E. McDougall & J. A. Ross (Eds.), *Proceedings of the twenty-sixth annual meeting of the North American Chapter of the International Group for the Psychology of Mathematics Education* (pp. 611–619). Toronto: OISE/UT.

Weiss, I. R., Banilower, E. R., McMahon, K. C., & Smith, P. S. (2001). *Report of the 2000 national survey of science and mathematics education*. Chapel Hill, NC: Horizon Research, Inc. Retrieved February 28, 2003, from http://2000survey.horizon-research.com/reports/status.php

## REFERENCES TO STATE-LEVEL GLE DOCUMENTS REVIEWED FOR THIS ANALYSIS*

Alabama Department of Education. (2003). *Alabama course of study: Mathematics*. Retrieved from http://www.alsde.edu/html/sections/documents.asp?section=54&sort=6&footer=sections

Alaska State Board of Education and Early Development. (2004). *Math performance standards (grade level expectations) for grades 3–10*. Retrieved from http://www.eed.state.ak.us/tls/assessment/GLEHome.html

Arizona Department of Education. (2003). *Arizona academic content standards—Mathematics (articulated by grade level)*. Retrieved from http://www.ade.az.gov/standards/math/articulated.asp

Arkansas Department of Education. (2004). *Arkansas mathematics curriculum framework*. Retrieved from http://arkedu.state.ar.us/curriculum/benchmarks.html#Math

California Department of Education. (2005). *Mathematics framework for California public schools, kindergarten through grade twelve*. Retrieved from http://www.cde.ca.gov/ci/ma/cf/

Colorado Department of Education. (2000). *Grade level expectations (examples)*. Document no longer available.

Connecticut State Department of Education. (2004). *Mathematics curriculum framework (draft)*. Retrieved from http://www.state.ct.us/sde/dtl/curriculum/ (Draft document no longer available—only final version).

Department of Defense Education Activity. (2004). *Mathematics curriculum content standards*. Retrieved from http://www.dodea.edu/instruction/curriculum/DoDEA_Content_Stand.htm

Florida Department of Education. (1996). *Sunshine State standards*. Retrieved from http://www.firn.edu/doe/curric/prek12/frame2.htm

Georgia Department of Education. (2004). *Georgia performance standards*. Retrieved from http://www.georgiastandards.org/math.aspx

Hawaii State Department of Education. (2004). *Framework and instructional guides—Grade level performance indicators*. Retrieved from http://standardstoolkit.k12.hi.us/sdb/database/index.jsp

Idaho State Department of Education. (2002). *Mathematics achievement standards*. Retrieved from http://www.boardofed.idaho.gov/saa/standards/math.asp

Idaho State Department of Education. (2005). *Idaho mathematics achievement standards*. Retrieved from http://www.sde.state.id.us/dept/standards.asp

Illinois State Board of Education. (2004). *Illinois mathematics assessment framework*. Retrieved from http://www.isbe.net/assessment/IAFindex.htm

Indiana Department of Education. (2000). *Indiana's academic standards for mathematics*. Retrieved from http://www.doe.state.in.us/standards/welcome2.html

Kansas State Department of Education. (2003). *Kansas curricular standards for mathematics*. Retrieved from http://www.ksde.org/outcomes/mathstd.html

Louisiana Department of Education. (2004). *Grade-level expectations*. Retrieved from http://www.doe.state.la.us/lde/ssa/1819.html

Maine Department of Education. (2004). *Grade-level expectations*. Retrieved from http://www.maine.gov/education/lsalt/gles.htm

Maryland State Department of Education. (2003). *Maryland voluntary state curriculum*. Retrieved from http://www.mdk12.org/instruction/curriculum/mathematics/index.html

Michigan Department of Education. (2004). *Michigan grade level content expectations*. Retrieved from http://www.michigan.gov/mde/0,1607,7-140-28753_33232---,00.html

Minnesota Department of Education. (2003). *Minnesota academic standards for mathematics*. Retrieved from http://education.state.mn.us/mde/Academic_Excellence/Academic_Standards_Curriculum_Instruction/Mathematics/index.html

Mississippi Department of Education. (1999). *Mississippi mathematics framework 2000*. Retrieved from http://www.mde.k12.ms.us/acad/id/curriculum/math/mathframe.htm

Missouri Department of Elementary and Secondary Education. (2004). *Mathematics grade-level expectations*. Retrieved from http://dese.mo.gov/divimprove/curriculum/GLE/MAgle.html

Nevada Department of Education. (2003). *Nevada content and performance standards*. Retrieved from http://www.doe.nv.gov/standards/standmath/math.html

New Hampshire Department of Education. (2004). *Mathematics local grade level expectations (K–8).* Retrieved from http://www.ed.state.nh.us/Education/doe/organization/curriculum/NECAP/GLEs.htm

New Jersey Department of Education. (2002). *New Jersey core curriculum content standards for mathematics.* Retrieved from http://www.njpep.org/standards/revised_standards/Math_newstandards/TOC.html

New Mexico Public Education Department. (2002). *Mathematics content standards, benchmarks, and performance standards.* Retrieved from http://www.nmlites.org/standards/math/index.html

New York State Education Department. (2005). *New York learning standards for mathematics.* Retrieved from http://www.emsc.nysed.gov/ciai/mst/mathstandards/mathcorepage.htm

North Dakota Department of Public Instruction. (2005). *Mathematics content and achievement standards.* Retrieved from http://www.dpi.state.nd.us/standard/content/math/index.shtm

Office of Academic Services of the District of Columbia Public Schools: Division of Standards and Curriculum. (2002). *Standards for learning and teaching.* Retrieved from http://www.k12.dc.us/dcps/curriculum/curriculum1.html

Office of Superintendent of Public Instruction—Washington (2004). *Mathematics K–10 grade level expectations: A new level of specificity.* Retrieved from http://www.k12.wa.us/curriculumInstruct/mathematics/default.aspx

Ohio Department of Education. (2001). *Academic content standards.* Retrieved from http://www.ode.state.oh.us/academic_content_standards/acsmath.asp

Oklahoma State Department of Education. (2002). *Priority academic student skills.* Retrieved from http://sde.state.ok.us/publ/pass.html

Oregon Department of Education. (2002). *Oregon grade level standards and K–2 foundations.* Retrieved from http://www.ode.state.or.us/teachlearn/subjects/mathematics/curriculum/

Public School of North Carolina State Board of Education. (2003). *Mathematics standard course of study and grade level competencies K–12.* Retrieved from http://www.ncpublicschools.org/curriculum/mathematics/scos/

Rhode Island Department of Elementary and Secondary Education. (2004). *NECAP and local mathematics grade level expectations.* Retrieved from http://www.ridoe.net/standards/gle/default.htm

South Carolina State Department of Education. (2001). *South Carolina mathematics curriculum standards 2000.* Retrieved from http://www.myscschools.com/offices/cso/mathematics/standards.htm

South Dakota Department of Education. (2004). *South Dakota mathematics content standards.* Retrieved from http://doe.sd.gov/contentstandards/math/standards.asp

State of Vermont Department of Education. (2004). *Grade expectations for Vermont's framework of standards and learning opportunities.* Retrieved from http://www.state.vt.us/educ/new/html/pgm_curriculum/mathematics.html

Tennessee State Board of Education. (2001). *Mathematics curriculum standards.* Retrieved from http://www.state.tn.us/education/ci/cistandards2001/math/cimath.htm

Texas Education Agency. (1998). *Texas essential skills and knowledge for mathematics.* Retrieved from http://www.tea.state.tx.us/rules/tac/chapter111/index.html

Utah State Office of Education. (2003). *Mathematics core curriculum.* Retrieved from http://www.usoe.k12.ut.us/curr/math/elem/default.htm

Virginia Department of Education. (2002). *Virginia mathematics standards of learning curriculum framework.* Retrieved from http://www.pen.k12.va.us/VDOE/Instruction/Math/math_framework.html

West Virginia Department of Education. (2003). *Mathematics content standards and objectives for West Virginia schools.* Retrieved from http://wvde.state.wv.us/csos/

Wyoming State Board of Education. (2003). *Wyoming mathematics content and performance standards.* Retrieved from http://www.k12.wy.us/SA/standards.asp

Note: *Web sites active as of January 1, 2006.

# APPENDICES

## Appendix A. Names and Publication Dates for State-Level Mathematics Curriculum Documents Analyzed for the Reports Presented in Chapters 2, 3, and 4[1]

| *Ch2* | *Ch3* | *Ch4* | *State* | *Document Title* | *Published* |
|---|---|---|---|---|---|
| X | X | X | Alabama | Alabama Course of Study: Mathematics | 2003 |
| X | X | X | Alaska | Math Performance Standards (Grade Level Expectations) for Grades 3–10 | 2004 |
| X | X | X | Arizona | Arizona Academic Content Standards–Mathematics (articulated by grade level) | 2003 |
| X | X | X | Arkansas | Arkansas Mathematics Curriculum Frameworks K–12 | 2004 |
| X | X | | California | Mathematics Framework for California Public Schools: K–12 | 2005 |
| X | X | X | Colorado | Grade Level Expectations (Examples) | 2000 |
| | X | | Connecticut | Mathematics Curriculum Framework (DRAFT) | 2004 |
| X | X | X | Dept. of Defense | Mathematics Curriculum Content Standards | 2004 |
| X | X | X | Dist. of Columbia | Standards for Teaching and Learning | 2002 |
| X | X | X | Florida | Sunshine State Standards | 1996 |
| X | X | X | Georgia | Georgia Performance Standards | 2004 |
| X | X | X | Hawaii | Framework and Instructional Guides--Grade Level Performance Indicators | 2004 |
| X | X | X | Idaho[2] | Idaho Mathematics Achievement Standards | 2005 |

(Appendix A continues on next page)

*The Intended Mathematics Curriculum as Represented in State-Level Curriculum Standards: Consensus or Confusion?* 123–126

**Appendix A. Continued**

| *Ch2* | *Ch3* | *Ch4* | *State* | *Document Title* | *Published* |
|---|---|---|---|---|---|
| | X | | Illinois | Illinois Mathematics Assessment Framework | 2004 |
| X | X | X | Indiana | Indiana's Academic Standards for Mathematics | 2000 |
| X | X | X | Kansas | Kansas Curricular Standards for Mathematics | 2003 |
| X | X | X | Louisiana | Grade Level Expectations | 2004 |
| X | | | Maine | Grade Level Expectations | 2004 |
| X | X | | Maryland | Maryland Voluntary State Curriculum | 2004 |
| X | X | X | Michigan | Michigan Grade Level Content Expectations (GLCE) | 2004 |
| X | X | X | Minnesota | Minnesota Academic Standards for Mathematics | 2003 |
| X | X | X | Mississippi | Mississippi Mathematics Framework 2000 | 1999 |
| X | X | X | Missouri | Mathematics Grade Level Expectations | 2004 |
| X | X | X | Nevada | Nevada Content & Performance Standards | 2003 |
| X | X | | New Hampshire[3] | Local Grade Level Expectations (K–8) (with RI) | 2004 |
| X | X | X | New Jersey | New Jersey Core Curriculum Content Standards for Mathematics | 2002 |
| X | X | X | New Mexico | Mathematics Content Standards, Benchmarks, and Performance Standards | 2002 |
| X | | | New York | New York Learning Standards for Mathematics | 2005 |
| X | X | X | North Carolina | Mathematics Standard Course of Study and Grade Level Competencies K–12 | 2003 |
| X | X | | North Dakota | Mathematics Content and Achievement Standards | 2005 |
| X | X | X | Ohio | Academic Content Standards K–12 Mathematics | 2001 |
| X | X | X | Oklahoma | Priority Academic Student Skills | 2002 |
| X | X | X | Oregon | Oregon Grade Level Standards and K–2 Foundations | 2002 |
| X | X | | Rhode Island[3] | NECAP and Local Mathematics Grade Level Expectations (K–8) (with NH) | 2004 |
| X | X | X | South Carolina | South Carolina Mathematics Curriculum Standards 2000 | 2001 |
| X | X | X | South Dakota | South Dakota Mathematics Content Standards | 2004 |
| X | X | X | Tennessee | Mathematics Curriculum Standards | 2001 |
| X | X | X | Texas | Texas Essential Knowledge and Skills for Mathematics | 1998 |
| X | X | X | Utah | Mathematics Core Curriculum | 2003 |
| X | X | X | Vermont | Grade Expectations for Vermont's Framework of Standards and Learning Opportunities | 2004 |
| X | X | X | Virginia | Virginia Mathematics Standards of Learning Curriculum Framework | 2002 |
| X | X | X | Washington | Mathematics K–10 Grade Level Expectations: A New Level of Specificity | 2004 |

(Appendix A. continues on next page)

**Appendix A. Continued**

| *Ch2* | *Ch3* | *Ch4* | *State* | *Document Title* | *Published* |
|---|---|---|---|---|---|
| X | X | X | West Virginia | Mathematics Content Standards and Objectives for West Virginia Schools | 2003 |
| X | X | X | Wyoming | Wyoming Mathematics Content and Performance Standards | 2003 |
| 42 | 42 | 35 | | | |

*Notes:* (1) As identified by a search of State Education Department Websites as of May 2005. Website link to each document can be found at http://mathcurriculumcenter.org/states.php
(2) For the analysis described in chapter 4, the 2002 Idaho document was used.
(3) New Hampshire and Rhode Island share a common document.

**Appendix B. Mean Number of Learning Expectations per Grade by Strand Within State GLE Documents[1]**

| *State* | *Num/Op* | *Alg* | *Geo* | *Meas.* | *Data/Prob* | *Total Mean per Grade* |
|---|---|---|---|---|---|---|
| AL | 11.4 | 3.5 | 5.6 | 5.1 | 3.4 | 29.0 |
| AK | 15.2 | 5.2 | 9.2 | 7.0 | 5.3 | 41.8 |
| AZ | 32.3 | 10.0 | 10.9 | 8.5 | 13.6 | 75.3 |
| AR | 24.0 | 9.9 | 8.8 | 13.5 | 7.3 | 63.4 |
| CA | 13.0 | 6.6 | 9.1 | | 5.6 | 34.3 |
| DoDEA | 9.8 | 7.8 | 6.1 | 6.9 | 7.6 | 38.1 |
| DC | 11.6 | 7.4 | 7.4 | 7.8 | 6.3 | 40.4 |
| FL | 27.8 | 12.6 | 12.4 | 17.4 | 14.5 | 84.6 |
| GA | 14.8 | 6.1 | 5.8 | 6.4 | 3.8 | 36.8 |
| HI | 15.0 | 7.1 | 10.0 | 9.0 | 14.1 | 55.3 |
| ID | 8.5 | 5.9 | 3.5 | 3.9 | 4.6 | 26.4 |
| IN | 16.6 | 7.4 | 6.6 | 8.6 | 3.4 | 42.6 |
| KS | 32.4 | 32.0 | 27.0 | | 15.5 | 106.9 |
| KY | 11.0 | 6.8 | 7.8 | | 8.4 | 34.0 |
| LA | 12.3 | 7.0 | 6.6 | 7.0 | 7.0 | 39.9 |
| ME | 4.3 | 3.3 | 2.3 | 2.2 | 2.7 | 14.8 |
| MD | 19.8 | 14.0 | 10.8 | 9.0 | 11.5 | 65.0 |
| MI | 19.8 | 5.0 | 7.1 | 7.0 | 3.6 | 42.5 |
| MN | 10.6 | 3.8 | 4.8 | 3.4 | 3.8 | 26.3 |
| MO | 9.0 | 6.4 | 6.3 | 5.4 | 4.8 | 31.8 |
| MS[2] | | | | | | 45.8 |
| NV | 8.3 | 5.3 | 6.6 | 4.8 | 3.0 | 27.9 |
| NH | 7.5 | 3.3 | 6.1 | | 5.4 | 22.3 |

(Appendix B continues on next page)

**Appendix B. Continued**

| *State* | *Num/Calc* | *ALG* | *Geo* | *Meas.* | *Data/Prob* | *Total Mean per Grade* |
|---|---|---|---|---|---|---|
| NJ | 22.5 | 14.7 | 26.3 | | 18.2 | 81.7 |
| NM | 19.8 | 16.3 | 13.3 | 9.0 | 19.0 | 77.3 |
| NY | 23.1 | 7.1 | 9.5 | 9.4 | 7.3 | 56.4 |
| NC | 10.9 | 4.0 | 4.5 | 3.0 | 3.5 | 25.9 |
| ND | 16.3 | 4.8 | 5.9 | 7.8 | 6.1 | 40.8 |
| OH | 13.0 | 8.4 | 6.6 | 7.1 | 9.5 | 44.6 |
| OK | 9.5 | 2.5 | 3.0 | 3.8 | 3.6 | 22.4 |
| OR | 18.0 | 8.4 | 10.1 | 10.3 | 7.6 | 54.4 |
| RI | 7.5 | 3.3 | 6.1 | | 5.4 | 22.3 |
| SC | 21.1 | 8.8 | 14.8 | 12.3 | 10.1 | 67.1 |
| SD | 8.4 | 8.6 | 5.1 | 5.6 | 3.8 | 31.5 |
| TN | 25.0 | 13.5 | 10.9 | 12.0 | 9.9 | 71.3 |
| TX | 10.6 | 5.4 | 5.4 | 4.4 | 4.8 | 30.5 |
| UT | 24.6 | 8.7 | 9.7 | 9.9 | 6.6 | 59.4 |
| VT | 6.8 | 3.6 | 6.6 | | 6.9 | 23.9 |
| VA | 11.9 | 3.8 | 4.6 | 7.6 | 4.4 | 32.3 |
| WA | 30.1 | 20.0 | 15.4 | 19.8 | 19.8 | 105.0 |
| WV | 12.3 | 7.3 | 7.1 | 8.0 | 4.6 | 39.3 |
| WY | 7.3 | 2.9 | 4.1 | 4.6 | 3.1 | 22.0 |
| Mean | 15.5 | 8.0 | 8.5 | 6.5 | 7.5 | 46.0 |

*Notes:* (1) Since state documents differ in their organizational structure, we counted GLEs according to the lowest hierarchical level. That is, if a state document included three levels of organization to convey learning expectations, we counted the number of individual statements at the lowest level, or in this case, third level. Retrieved from State Education Department Websites as of May 2005.
(2) GLEs in the MS document are integrated, not sorted by strand. Therefore, only the mean per grade level is shown in this table.

Printed in the United States
71196LV00001B/48